ORIOMO PAPUANS

ORIOMO PAPUANS

Ecology of Sago-Eaters in Lowland Papua

Ryutaro Ohtsuka

UNIVERSITY OF TOKYO PRESS

Publication of this book was supported by a grant-in-aid from the Ministry of Education, Science and Culture of Japan.

UTP 3045-66452-5149
ISBN 0-86008-327-6

Printed in Japan.

Contents

Preface

It was in 1966 that Dr. Hitoshi Watanabe and I began framing a research project aimed at a functional-ecological understanding of the adaptive mechanisms of a contemporary human population under a subsistence regime. At that time, Dr. Watanabe was Lecturer in Anthropology, Faculty of Science, University of Tokyo (presently, Professor of Anthropology at Hokkaido University), and I was in the final stage of an undergraduate course in anthropology at the same university. From the outset of the project, I collaborated as a research assistant. The essentials of our methodological strategy for the fieldwork consisted of an ecological or ethnoarchaeological approach to the material conditions of man's life and an ecological or activity-centered approach to the man-environment relationship. The former approach was mainly pursued by Dr. Watanabe while the latter was my own principal tool in collecting data in the field and, thus, in shaping this monograph.

The selection of the area for fieldwork was a major problem. Having surveyed the literature, we were drawn to the low, swampy area in the middle, southern part of the main island of New Guinea facing the Australian continent because, in spite of a great interest in different subsistence systems between the hunting-gathering Australians and the horticultural Papua New Guineans, the people inhabiting the area in question had not been extensively researched by human ecologists, anthropologists or archaeologists, and simply because the area was one of the lesser-known zones in the South Pacific. Communication with many authorities confirmed the area as one of scientific interest and importance. We finally selected the Oriomo Plateau, which is a big deltaic area surrounded by the River Fly on the east, the River Digul (Digoel) in Irian Jaya on the west and the Torres Strait on the south. Those who gave us especially helpful suggestions and encouragement in our research project were Jack Golson and D. J. Mulvaney, Professors at the Australian National University in Canberra, and Ron Crocombe, Professor at the New Guinea Research Unit of the Australian National University in Port Moresby (presently, Professor at the University of the South Pacific).

In spite of our preparatory process, i.e., during communication with scholars in Australia and Papua New Guinea, and during perusal of exploratory and ethnographic monographs about the Oriomo Plateau and its surroundings (Beaver 1920; Landtman 1927; Williams 1936; Archbold and Rand 1940), our picture of present-day conditions in the area and the contemporary lifestyle of the inhabitants remained unclear. The most manifest reason is derived from the lack of field investigation in the area during the last several decades. It was said that the swampy nature of the area had kept researchers away. Although an Australian scholar informed us that the swampy condition of the area had dissuaded them, it occurred to me that Japanese researchers' possibly higher adaptability, in a physiological sense, to swampy land would be of advantage.

Late in 1966 an announcement arrived stating that our proposed research project, which was planned to begin in July of 1967, had won a grant from the Wenner-Gren Foundation for Anthropological Research (Director, Dr. Lita Osmundsen). Until the actual outset of fieldwork, my preparatory efforts were mainly devoted to gaining a technical acquaintance with data collecting, especially in the area of concrete descriptions of daily activities on a time-space dimension. I visited, at sporadic intervals, several rural communities in Japan to observe people involved in agriculture, fishing and other occupational activities. In addition, my observations extended to such non-human primates as chimpanzees in an open-field zoo in Tokyo.

In June of 1967, when I had just entered the Masters Course in Anthropology at the University of Tokyo, I left Japan by sea for Papua New Guinea (at that time, Papua and New Guinea were known as the External Territory of Australia).

According to our plan, Dr. Watanabe would have arrived by air ahead of me in Port Moresby and would meet me in July, but a sudden stomach disease forced him into a hospital bed while I was on the sea voyage. After I arrived in Port Moresby, I consulted with Dr. Watanabe and Dr. Crocombe, Professor of the New Guinea Research Unit there, and a decision was reached that, accompanied by Mr. Robin Hide, the then Research Assistant of Professor Crocombe, I would carry out a short, preliminary trip in the target area, while the planned fieldwork would be postponed until Dr. Watanabe's recovery. The resulting two-week glance at the area influenced to a great extent, directly or indirectly, the reworking of our (especially my) fieldwork design, which was resuscitated in 1971.

The passage of four years before the recommencement of fieldwork in 1971 provided me with a chance to continue data collecting in the field and to perform the associated analyses. I did fieldwork among several rural communities in Japan, and my study of the hand-line fishermen, from which my M.S. Thesis and several relevant articles emerged (Ohtsuka 1970, 1971, 1972, 1977b), enhanced my research strategy. Concretely, my special attention in such fieldwork focused on a detailed analysis of daily activities, especially subsistence activities, as a temporal and spatial structure system. I further considered the applicability of these observations to the Oriomo study.

In 1970, before Dr. Watanabe's complete recovery, I moved as an instructor to the Department of Human Ecology, School of Health Sciences, Faculty of Medicine, University of Tokyo, to which I still belong. Fresh stimuli from the members of the new department gave me a number of hints in structuring research themes and associated methodologies. With support from the staff of the Human Ecology Department, especially from Professors H. Katsunuma and T. Suzuki, I was engaged in fieldwork, in collaboration with Dr. Watanabe, among the Oriomo Papuans for eight months duration between June 1971 and March 1972. Although Dr. Watanabe's stay in the field was restricted to the beginning six weeks for fear of his health, his study produced *Bow and Arrow Census in a West Papuan Community: A New Field of Functional-Ecological Study* (Watanabe 1975).

After an eight-year discontinuation following the 1971–72

survey, our new research project among the Oriomo Papuans commenced in 1980. The new project, the title of which is "Human adaptability in lowlands of the South Pacific," is headed by Dr. T. Suzuki, Professor of Human Ecology at the University of Tokyo, and is financially supported by the Japanese Ministry of Education, Science and Culture. The new project intended to focus upon the adaptive mechanisms of a human population as a surviving unit through generations, dealing with all the people of one linguistic group, including the villagers of Wonie, the target of this monograph. This ongoing project consisted of a three-month preliminary survey carried out by two of the members, myself and Mr. T. Kawabe of the same department, from June to September 1980, and a main survey from 1981 to 1982 by six researchers, five Japanese and one Papua New Guinean. While the data gathered in this new project is outside the scope of this monograph, some firsthand information about the population will contribute to a broader understanding of the Gidra-speaking group as a whole. In addition, a part of the analyzed data for the nutrient composition of food samples gathered in 1980 and 1981–82 will be used for a more reasonable assessment of the foods exploited by the Wonie villagers in 1971–72.

Data from the present fieldwork appear in some other briefer papers which have already been published (mainly, Ohtsuka 1977a, 1977c). The contents of these papers partially overlap with this volume.

Complicated statistical analyses in this study were made using the HITAC 8700/8800 system in the University of Tokyo Computer Center. For the computations for cluster analysis and partial correlation analysis, the SPSS (Statistical Package for Social Sciences) program package (Nie, Hull, Jenkins, Steinbrenner and Bent 1975; Miyake and Yamamoto 1976; Miyake, Nakano, Mizuno and Yamamoto 1976) was applied.

Acknowledgments

The people of Wonie among whom I stayed and worked for this study generously accepted me not only as a researcher but also as their friend. Their continuous friendliness to me was indispensable to my fieldwork. A number of people inhabiting the neighboring villages were also helpful.

Financial support for the fieldwork was provided by a Wenner-Gren Foundation for Anthropological Research grant (#21125) awarded to Dr. Hitoshi Watanabe of the University of Tokyo (presently, Professor at Hokkaido University), with whom I cooperated in the field and to whom I have been most indebted for performing the fieldwork and for making up this monograph.

Thanks are also due to Professors J. Golson and D.J. Mulvaney and to many other researchers of the Australian National University in Canberra for their suggestions regarding the selection of the Oriomo as our study area; to the New Guinea Research Unit of the Australian National University in Port Moresby, especially to Professor R. Crocombe and Dr. R. Hide in 1967 and Dr. M. Ward in 1971–72, and to Professor R. Bulmer of the University of Papua New Guinea in 1971 for arranging fieldwork; and to the Research School of Pacific Studies of the Australian National University for providing me with the opportunity to gather information there concerning this study in 1974.

Botanical specimens were identified by Mr. J. S. Womersley, Mr. E. E. Henty, Mr. M. Galore and their colleagues of the Department of Botany (the Office of Forests) in Lae, and the identification of several kinds of animals was due to Mr. F. Parker, who then worked as a patrol officer of the Western

District in Daru. I am also grateful to Mr. Parker for transportation to the study area and other facilities in Daru.

Finally, my thanks extend to the members of the research project of "Human adaptability in lowlands of the South Pacific": Dr. T. Suzuki (Project leader), Mr. T. Akimichi, Mr. T. Kawabe, Mr. T. Inaoka and Mr. B. Laba, for their cooperation in the field in 1980 and in 1981–82 and for some of the data used in this monograph; and to the staff and colleagues of the Department of Human Ecology, University of Tokyo, for their continuous support.

1. Introduction

1. Peoples and Places of the Oriomo

Geography, Fauna and Flora

The Oriomo Plateau is located in the middle, southern part of the main island of New Guinea, between approximately 138°E and 143°E in longitude, and between 7°S and 9°S in latitude (Fig. 1, Photo 1). The plateau is 450 km in length, stretching from the River Digul (Digoel) in Irian Jaya (West Irian) to the Fly River delta, and averages from 80 to 125 km in width. Administratively, the area of its eastern half belongs to the Western Province of Papua New Guinea, but in my survey period of 1971 to 1972 it was a part of the Western District of Papua, known as the Australian External Territory. The Western Province (District) has been the most thinly-populated in Papua New Guinea, and Daru, the capital of the district (province), possessed a population of less than 4,000, of which one-twentieth was non-indigenous in my study time.

The deltaic Oriomo Plateau rises some 60 m above the surrounding marshlands at its highest point. Geomorphologically, it is characterized by a low swampy plain, with a low ridge, called the Morehead Ridge (Blake 1971), approximately a dozen meters above the surroundings, extending west to east in the central zone. In the eastern inland area, including the study village of Wonie, complicated networks of small creeks form the upper parts of the rivers, i.e., the Bituri running north to the Fly River and the Oriomo and Binaturi (or Ume) running south to the Torres Strait.

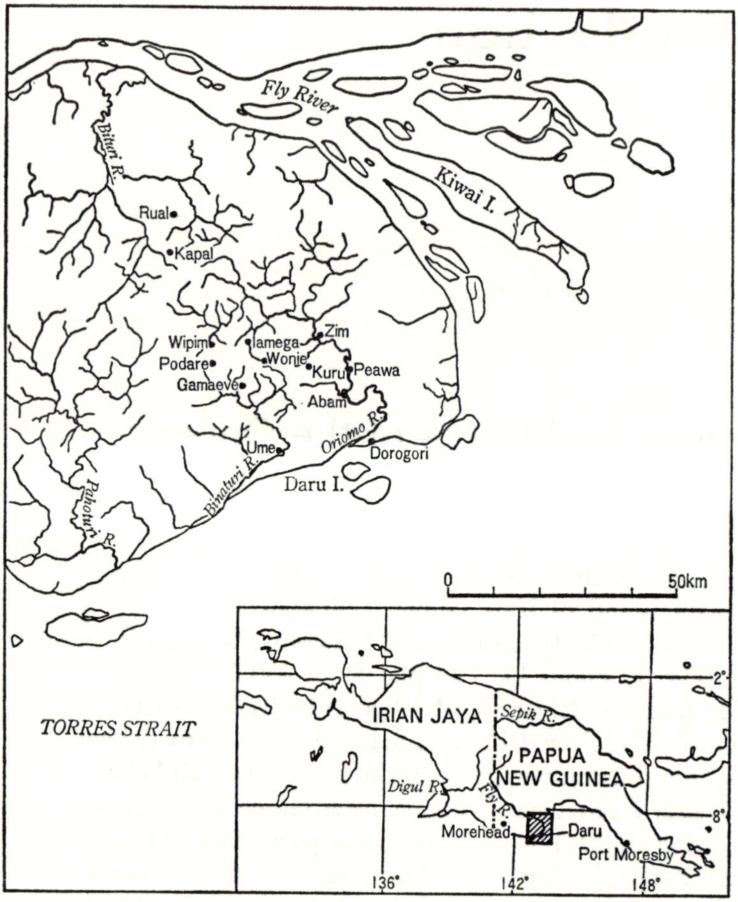

Fig. 1. The Eastern End of the Oriomo Plateau, Papua New Guinea, Showing Position of Wonie Village and Twelve Other Gidra-speaking Villages.

According to meteorological records in Daru (McAlpine 1971), where the nearest observatory is located, the average monthly temperature fluctuates only from 25°C to 28°C (Table 1). My own measurements demonstrated that the daily fluctuation in temperature on a day of fine weather had a range of about 10°C, i.e., 30°C or a little higher at the maximum and approximately 20°C at the minimum, while that on a rainy day ranged within several degrees around 25°C. The average annual rainfall reaches 2,000 mm, with about 80 percent falling during the wet season from December to May, while the four-year records in Morehead, approximately 100 km

Table 1. Meteorological records in Daru and Morehead

Temperature (°C) in Daru: 1894–1967

	Jan	Feb	Mar	Apr	May	Jun	Jul	Aug	Sep	Oct	Nov	Dec	average
mean of maximum	32.0	31.8	31.4	30.6	30.2	29.1	28.4	28.7	29.5	31.0	32.2	32.5	30.6
mean	27.7	27.6	27.3	26.9	26.8	25.9	25.3	25.6	26.2	27.2	28.0	28.0	26.9
mean of minimum	23.5	23.3	23.1	23.3	23.4	22.7	22.3	22.5	22.9	23.4	23.8	23.6	23.2

Rainfall (mm)

location	period	Jan	Feb	Mar	Apr	May	Jun	Jul	Aug	Sep	Oct	Nov	Dec	total
Daru	1894–1967	278.9	253.2	331.7	318.3	224.0	99.1	90.9	53.4	41.9	56.1	111.8	207.5	2,067.1
Morehead	1964–1967	302.3	294.1	323.6	138.4	193.0	69.1	59.7	76.7	35.1	46.2	135.4	200.7	1,879.3

west of the study area, show an average of about 1,900 mm, with a similar seasonal cycle (Table 1). The timing of rainfall is unpredictable in both the dry and wet seasons. No specific time of day when rainfall concentrated, nor any clear time interval of rainfall in any season, was found. According to my impression, heavy showers predominated over soft, continuous rainfall in terms of frequency not only in the wet season but also in the dry season. In the wet season the creeks frequently overflow into savanna and woodland, while in the dry season they form segregated waterpools and, eventually, dried-up depressions. According to my observations, the paths and even the village yard were sometimes inundated in February and March 1972. On the contrary, the supply of drinking water became marginal in the middle of the dry season. The situation was more severe in our 1980 survey period; in September the water-supplying creek of the Wonie people, several minutes' walk from the village, contained no water. Moreover, the people of Kuru village, eastward adjacent to Wonie, could not obtain water from the area surrounding the village settlement and had to abandon it temporarily.

Another feature of the natural environment in the Oriomo Plateau is a lack of stones other than a small number of pisolitic laterites. Although no prehistoric study has been made in the area, stone tools such as axes and grinders might have existed since antiquity, according to the contemporary people. I found several of these tools outside the village settlement, but their origin is unknown. These stone implements are used for limited and specific purposes, like crushing wild nuts. For implemental conditions, except for the stone axes (which have been replaced by steel axes), the present-day villagers' knowledge suggests a high possibility that in olden times freshwater and brackish-water shells (at least, *Virgus* sp. and some other unidentified species), a variety of bamboos (*Bambusa vulgaris*, etc.) and the hard woods of trees were used as implements in spite of their infrequent use at present.

The dominant types of vegetation in the eastern region of the plateau, including the study region, are savanna and monsoon forests, while riverine forests and small patches of savanna woodland occur. However, the natural vegetation in the area has been disturbed to a considerable degree by human activities like shifting cultivation practices, grass-burning for hunting and other unspecified activities (cf. Paijmans 1971). This area produces eleven ethnotaxonomic categories of vegetation. Of these, four basic types are woodland called *bua*, thickly wooded savanna *kuur*, thinly wooded savanna *yap* and grassland without tall trees *tupi*. Small grassland in *bua* is distinguished as *kuk*, while small patched woodland in *yap* is called *buayap*. However, *buayap* is not clearly discriminated from *kuur* even among the villagers. A puzzling category called *uotomibua* indicates savanna in the process of succession from grassland to woodland, as determined by the people's observation; thus this cannot be distinguished from *yap* by researchers from outside the region. Furthermore, riverine forests along very small creeks are called *posenibua*, and along big creeks, *bojogbua*, while those along big and permanently-running rivers like the Oriomo and Binaturi are called *kulumbua*. The last category is *biit*, which indicates a permanently inundated swamp. The classification of vegetation is well known by the people, but the vegetation is fairly changeable because of the speedy regrowth of trees in *tupi* or *yap*. For example, an area of some square km, which had been *tupi* when the living aged men were young adolescents, passed from *uotomibua* to *kuur* during my study time. This volume will refer to woodland (*bua*), savanna woodland (*kuur* and/or *buayap*), savanna (*yap*) and riverine forest of *bojogbua* since the remainder occupy negligible areas in the study region and are not specifically useful for the people's activities, such as procurement of food resources.

There is a wide variety of tree species in various vegetations. According to my botanically unskilled observations, the clearly dominant trees in the savanna are *Melaleuca* spp., and their full-grown height is 10–15 m on the average. There is no species which predominates throughout the forest, and the canopy of the woodlands reaches 30 m or more, while that of other forest-type vegetations differs from place to place. Except for those trees specifically used by the people, the natural stands were not systematically identified in my study, although the results of the dendrological research in the Morehead-Kiunga area 100 km west of the study region (Paijmans 1971) may be applicable; his results are, in summarized form, shown in Appendix 1. The grasses which predominate over the savanna, as far as observed in the study area, were collected for identification by botanists in the Department of Forests in Lae. The most common are *Imperata cylindrica* and *Pseudopogonatherum irritans*, and the remainder include four species of the family Gramineae, i.e., *Coelorachis rottboellioides*, *Eremochloa ciliaris*, *Ischaemum barbatum* and *Eleusine indica*; three species of Cyperaceae, i.e., *Rhynchospora rubra*, *Fimbristylis insignis* and *Cyperus sphacelatus*; and one species of Eriocaulaceae, i.e., *Eriocaulon australe*. The height of grasses changes according to the burning done by the people. In their natural state the predominant grasses of *Imperata cylindrica* and *Pseudopogonatherum irritans* grow about 2–3 m in height.

With regard to fauna, New Guinea belongs to the Australian zoogeographical region; hence, the indigenous mammalians inhabiting this area are marsupials and monotremes. However, dogs, pigs and rats were introduced to New Guinea in olden times. The presence of a non-indigenous animal, the pig, is worthy of special note, since it has been regarded as an element in horticultural systems throughout the Pacific (Golson 1972). S. Bulmer (1966), basing her results on her excavations in the Highlands of New Guinea, established that pigs (*Sus scrofa*) inhabited the area at least as early as B.C. 3000, while new evidence traces their existence back a further 1,000 years (White 1971b, 1972).

Marsupials abundant in the Oriomo area are three species of wallabies (*Wallabia agilis* called grass-wallaby in this volume; *Tylogale* sp., called scrub-wallaby; *Dorcopsis veterum*, called bush-wallaby) and two species of bandicoots (*Echymipera* spp.). Several species of possum (*Phalanger* spp., etc.) are seen infrequently. Echidna (*Tachyglossus aculeatus*), the only species of monotreme in this area, was well known by the people, but was not witnessed by me in the 1971–72 period; in 1981 I saw one such animal which had been killed by a hunter of the Wonie village. While there were more than one hundred vernacular names for the birds gathered in my fieldwork, the important game animals are first the cassowary (*Casuarius casuarius*) belonging to the Rattiae, next a megapode (genus *Megapodius*), and finally various species of pigeons, doves and hornbills. A variety of small-sized birds, for example, parrots and lories and even a kind of bird of paradise (*Paradisea raggiana*), inhabit the area but are not usually killed by the people.

Reptiles inhabiting the area, identified or at least distinguished, consisted of a species of crocodile (*Crocodilus porosus*), eight species of monitor lizard (*Varanus* spp., etc.), seven species of lizard (*Egernia* sp, *Tiliqua* sp., etc.) and two species of tortoise (*Chelodina* sp. and *Emydura* sp.). Those captured and eaten, though not frequently, include most species of the monitor, some species of the snake, and both species of the tortoise. Judging from the fishing catches, at least twenty or thirty fish species inhabit the creeks, although these were not ichthyologically identified. Of a variety of insects, two species of tree ant (*Oecophylla virescens* and another one belonging to the same genus but unidentified), three kinds of grub (or larvae) of the family Cerambycidae, one of which is perhaps *Rhynchophorus schach* Olive (cf. Kimura 1979), and bee honey are used as food.

Animals dangerous to human life include the pig and cassowary, both of which are major game animals in the area. According to R. Bulmer (1969), throughout New Guinea only the above two kinds of creatures grow as large as man, and they are the only ones which are dangerous if attacked, being capable of maiming or killing a hunter with their tusks or claws, respectively. Besides these, however, crocodiles and several species of poisonous snake, which are sometimes hunted, are also dangerous. During my stay in the area, an adolescent male of Wipim, a village adjacent to Wonie, was killed by a Papuan black snake (*Pseudechis papuanus*), and a number of episodes suggest dangerous snake bites. A variety of noxious insects and microorganisms may have taken the lives of some of the people, especially the infants, but I did not determine a medically causal interrelationship.

Gidra Linguistic Group

The Trans-Fly area, including the eastern half of the Oriomo Plateau, is occupied by a number of linguistically rather diverse groups of people. Habitation is sparse except for the town of Daru. The language spoken by the subject people in Wonie is generally called Gidra, but the speakers of this language believe that Gidra was named by their coastal neighbors and they prefer the name Wipitungam, which originally implied *man* in their language, to Gidra. In this volume, however, I will use Gidra due to its more widespread acceptance.

In olden times, the peoples in the inland of the eastern Oriomo Plateau were called "bush men" by their coastal neighbors who lived, for example, in the villages of Mawata and Turituri (Tureture) (Beaver 1920: 79), and their languages might not have been fully distinguished from one another by researchers and perhaps by the coastal peoples. In a series of reports by the Cambridge Anthropological Expedition to Torres Strait, which was headed by A.C. Haddon, Ray (1907: 284–412) identified the languages not only of the islanders of the Torres Strait but also of their neighbors. He reported that the language spoken by the inland people, who might include the Gidra people, was called Jibu. Comparing a number of words recorded by Ray with contemporary Gidra words, Jibu might have implied Gidra, although more than half of the words were not known by the contemporary people. Thus, it is likely that the Jibu words collected by Ray

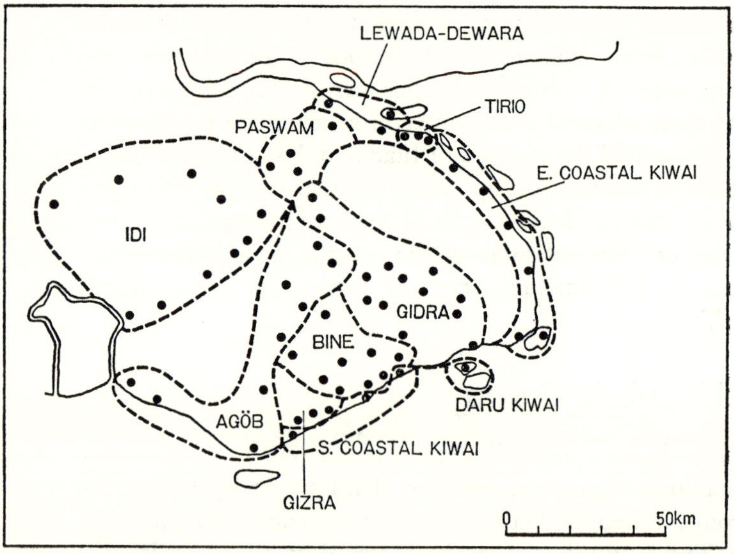

Fig. 2. Linguistic Boundaries around the Study Region.
Location of villages involved in three linguistic groups of the Eastern Trans-Fly Family (Gidra, Bine, Gizra) and the adjacent eight linguistic groups are plotted. Source: Wurm (1971b), whose original map covers a wider area.

were taken from a mixture of sources from several different languages of the inland peoples.

Thanks to recent linguistic investigations, Gidra was first identified as one of the three languages, i.e., Gidra, Bine and Gizra, of the Oriomo River Family (Voorhoeve 1968, 1970; Wurm 1971a). Since then Wurm (1971b) has discovered a close linkage between these three languages and Miriam, a language whose users inhabit the Murray Islands in the eastern Torres Strait, some 60 km southeast of Daru. As a result, the name of the Oriomo River Family was changed to the Eastern Trans-Fly Family, one of the families belonging to the Trans-Fly Stock (Fig. 2). The lexical interrelationship of the four languages of this family is relatively low. The lexicostatistical percentages obtained by using a 200-item basic vocabulary list (Wurm 1971b) are shown in Table 2. Wurm commented on the low percentage, i.e., 26%, for Gidra-Miriam: "Though this percentage is below the level diagnostic for a family-level relationship on the basis of a 200-item vocabulary list, the presence of a relationship chain of persistently above 28% which links Gidra with Miriam through Gizra makes it possible to regard both Gidra and Miriam as members of the same family." In spite of the linguistic connection, the Gidra-speakers' verbal communication with adjacent linguistic groups was possible using either of the two languages only between a limited number of persons, especially males, who had close relatives from other groups and/or had specific experiences such as frequent mutual visits.

Based on Wurm's (1971b) approximate estimations for the population of each linguistic group, the Trans-Fly Stock is comprised of 34 languages of five families involving 32,060 people. The population per linguistic group ranges from less than 100 to nearly 10,000, the majority having from several

Table 2. Lexicostatistical percentages of four languages of the Eastern Trans-Fly Family

	Bine	Gizra	Miriam
Gidra	42	34	26
Bine		38	28
Gizra			37

Source: Wurm (1971b).

hundred to 2,000 speakers. In the Eastern Trans-Fly Family, the size of each linguistic population was 1,600, 1,550, 600 and 700 respectively for Gidra, Bine, Gizra and Miriam. The time of estimating the above population numbers was not clearly described in Wurm's report, but it may be reasonable to suppose that the figures were based on information from the 1960's. Our investigation of the *de facto* population of the thirteen Gidra-speaking villages in 1980 showed a population of a little more than 1,800.

Geographically, the thirteen Gidra-speaking villages are situated in an oval-shaped territory of some 4,000 square km of land, extending southeast and northwest (see Figs. 1 and 2). Of these, one village called Dorogori (or Mawoto) is located along the coast facing Daru across a small strait. Three villages, Abam, Peawa (or Woigi) and Zim (or Wuroi), are on the river bank of the Oriomo, one village called Ume is on the bank of the Binaturi River (sometimes called the Ume River); and the remainder, i.e., Kuru, Wonie, Gamaeve, Podare, Wipim, Iamega, Kapal and Rual, are inland villages, while Rual and Kapal are situated fairly close to the tributaries of the Bituri River in the north. The geographical settings condition the way of living to some extent: for example, the people of the coastal village, Dorogori, depend on fishing for their protein supply, in contrast to the predominance of subsistence hunting among the inland people; coastal and riverine villagers travel mainly by canoe while inland villagers travel exclusively on foot.

The Gidra linguistic group does not have a traditional chief, and mutual communication among the people of the thirteen villages is not frequent. For instance, before our study in 1980 none of the Wonie villagers had had the experience of having visited all the thirteen villages. Nonetheless, the people of the Gidra group have kept the feeling that they share a certain unity, which may manifest itself not only linguistically but also socioculturally and genealogically. The sociocultural unity is manifestly seen in the totemic clan system. Each of the Gidra members belongs to a totem clan which is organized through the male line, and the Gidra territory is latently owned in various portions by the clans. The number of clans amounts to more than thirty today, although some others are said to have disappeared in the past. It is certain that the clan system has a long antiquity, but historical discernment is hardly possible and only a mythical story of the origin of the Dorem clan, which is predominant in Wonie, was gathered (Appendix 2). In terms of village organization in relation to the clan system, each contemporary village consists of persons, males and females, who belong to from several to more than ten different clans.

The Gidra people are basically endogamous, although in the peripheral regions of their territory the occurrence of exogamy with the people from the adjacent linguistic groups is relatively high. The recent expansion of the living sphere associated with temporary and (semi-)permanent migrations to the town of Daru and other urban areas has accelerated inter-linguistic group marriages. Up to 1972, all Wonie villagers had married within the Gidra group, but three exogamous marriages had taken place by 1980. Nonetheless, according to our 1981–82 survey, the proportion of inmigrants among the total contemporary population of the thirteen Gidra-speaking villages was still only 2.7%.

The above-mentioned characteristics demonstrate that the Gidra people form a population defined biologically, specifically ecologically and genetically, as "a collective group of organisms of the same species occupying a particular space" (Odum 1971: 162), and alternately defined as "a reproductive community (biotic community) of sexual and cross-fertilizing individuals which share a common gene pool" (Dobzhansky 1968). Along with the inter-village variety of geographical conditions, the diversified adaptive mechanisms of the Gidra people attract population ecological interest. In fact, our research project for the whole Gidra people in 1980 and 1981–82 focused upon this, although the present volume will treat human adaptation in an inland village, Wonie.

According to two aged men of Wonie, whose age was approximated to be 60 years old in 1971–72 and both of whom died before 1980, when they were young boys small-scale warfare had broken out between the Gidra and adjacent groups, though infrequently. While the Gidra group has no traditional fighting leader nor fighting organization, their fathers and/or forefathers used arrows, the heads of which were made of femurs of feral pigs, for killing the enemy.

Fighting within the Gidra group occurred rather frequently and still occurs infrequently although, at least in recent years, the use of such dangerous weapons as bow and arrow has vanished. The major causes of inter-linguistic group warfare were said to derive from land tenure and the capture of women, while those of fighting among the Gidra are connected with cumulated tension between the clans on various trifling matters. For the last several decades, no warfare or severe conflict involving the sacrifice of life has occurred between the linguistic groups nor within the Gidra group. At the present time, the Gidra people live in peace with their neighbors; more precisely speaking, hostile or mutual aid interests among the local groups, either inter-linguistically or intra-linguistically, scarcely exist, and thus the present people make their livelihood fairly independently as village groups.

Social Structure

There are more than thirty totemic clans in the Gidra population. The people believe that each clan is formally headed by a *mopruga,* normally the oldest male, but his actual rights and duties are not obvious. His special role is evident only in settling quarrels which break out among the villagers and in making calls of inquiry and condolence to other villages, although, in my view, these reflect his age more than his role as a clan leader *per se.*

Each clan possesses a main totem called *kakiena* and several subordinate totems called *iena.* The clans are classified into two groups, i.e., moieties, forming an exogamous relationship. Although the people have no name for the moieties, in this volume they are called Dorem moiety and Dantungam moiety after the names of the respective predominant clan. In the 1980 survey we identified 34 clans, 26 of which had already been identified in the 1971–72 survey period. The classification of the clans into two moieties and the totems of each clan, which are shown in Table 3, are based on the results of the 1971–72 study; however, not all of their *kakiena* and especially of their *iena* were made clear because the people tended to hesitate about declaring their own totems. The totems are mostly animals such as grass-wallaby (called *seba* locally), one species of bandicoot (*ruei*), and several species of snake, lizard,

Table 3. Clans of the Gidra group and their *kakiena* and *iena*

Dorem moiety

clan name	*kakiena*	*iena*
Dorem	seba (grass-wallaby)	ruei (bandicoot), worum (pig), muya (monitor)
Wammopam	kiter (snake)	seba
Rekam	seba	tatasi (bird)
Kameam	bawar (flying-fox)	
Masam	muya	
Sebaiena	seba	
Mognayam	seba	
Damuram	piako (monitor)	
Daream	bigad (flying-fox)	
Dugitungam	woisem (? man-like animal)	
Morob		
Wipmam		
Ijobora		
Midam		

Dantungam moiety

clan name	*kakiena*	*iena*
Dantungam	grem (snake)	raya (bird), nayugrem (eel)
Gameam	grem	diwar (catfish), gui (bird), gigi (cassowary)
Gaepiena	gaep (bird)	
Rayaiena	raya	
Moram	durupia (pandanus)	
Idara		
Yokunjam		
Tonyukakam		
Dumaram		
Wesitungam		
Sokutungam		
Yapdumaram		

bird and fish. According to some informants, *seba* is a *kakiena*, or one of the *iena* of all clans belonging to the Dorem moiety, while *grem* (one species of python snake) is a *kakiena* or one of the *iena* of all clans of the Dantungam moiety.

Each clan owns a portion of the Gidra land, and the boundaries between clan-lands are established throughout the Gidra territory. In contrast to the strict boundaries between clan-lands, the boundaries between the village-lands are loose, the people themselves dividing, in actual practice, the

Gidra territory into village-lands according to the distance between the village settlements; and conflicts over village boundaries do not occur. The area of a clan-land is smaller than that of a village-land; the former is usually a half or a third of the latter, so that a single village-land contains several clan-lands, partly or as a whole. The areas which are remote from any village settlements are utilized only for sago and coconut exploitation, and each of the sago and coconut groves in such locations is visited only a few times per year. Since a number of persons inhabit villages which are out of their own clan-lands, it may be natural to exploit any clan-lands within the village-land where they live. Actually, the people procure natural resources, make slash-and-burn gardens, and even plant sago shoots and coconut seedlings in their village-land, while their major sago and coconut groves are usually concentrated in their clan-land.

Exogamy between the two moieties is still observed. Of the collected records of 73 marriages which took place before 1972 among the peoples of my study village and their relatives, alive or dead, there was only one exception—the latest case. The people say that ideally two marriages should be organized simultaneously by means of an exchange of women. Of the 73 marriages, however, 26 sets or 52 marriages followed this system but the other 21 occurred independently. Furthermore, as shown in Table 4, 60 percent of the 52 exchanged women were neither sisters nor daughters of their bridegroom's marriage-exchange partners. The people recognize such a case as follows—the girl to be a bride is first presented to her bridegroom's partner by her brother or father and is then exchanged. The traditional marriage system does not demand a marriage payment, even in the non-exchange cases. These marriage customs may typify the people's generosity, which accords with

Table 4. Fify-two exchanged brides classified according to relation to the bridegroom's marriage-exchange partner

	no.
sister	14
daughter	7
others	
belonging to same clan	18
belonging to other clans	13

"generalized reciprocity" as defined and elaborated by Sahlins (1965) and Service (1966), that is, a form of exchange based on the assumption that returns will balance out in the long run. However, recent acculturation has introduced the idea of bride price, and in fact several recent non-exchange marriages in Wonie followed this new system although the payments were small.

Monogamous and polygynous families coexist among the Gidra society. The first marriage of a male leads to the establishment of a new household. Thus, the household involves a head and his wife or wives and, if any, their unmarried children. Monogamous couples predominate over the polygynous in number. Very few of the males who have reached marriageable age do not marry, and all females in the reproductive age marry. Those who have become widows or widowers usually remarry as long as they are in the reproductive stage. In Wonie, during my study period of 1971–72, fifteen households were monogamous, three polygynous, and one was a widow's. The existence of some polygynous families is mainly due to an approximately ten-year difference in age between sexes at initial marriage, which implies a sexual difference in the marital period. The decision to make a polygynous family is that of the husband, and the major reason is usually that of the sterility or an insufficient number of live births of his wife or wives. The decision is independent of economic status simply because there is a negligible difference in the possession of wealth in this less-urbanized society.

About half of the marriages take place between men and women from the same village. In cases of spouses from different villages, virilocal residence is predominant although uxorilocal residence is also found. Generally speaking, uxorilocal residence lasts for a short duration, i.e., a couple of years or so, until movement to the husband's birth village. A household possesses a dwelling house in the village. The dwelling house of a polygynous family is divided into sections for the number of wives, and, though plural wives are not formally ranked, the husband usually spends much of his time in a section specific to one of his wives.

The social organization is patrilineal. Infants of either sex belong to their father's clan and are never transferred if they are adopted. Among the Gidra group, so-called "adoption" of children when they are just infants, or sometimes even before birth, is fairly popular, and adopted boys and girls are called *okatibuga* and *okatingamugai*, respectively. In addition to this type of adoption, boys and girls are sometimes adopted when one or both of their parents have died, and they are also called *okatibuga* or *okatingamugai*. Adoption occurs through the volition of the adopter rather than through that of real parents. In a usual case, the adopter has a consanguineous or affinal relation with one of the parents of the child and/or inhabits the village where the child was born, although neither consanguineous, affinal nor residential relation is necessarily required. In spite of the fairly large number of adoptions, only a few cases lead to the transference of the child to the household of his or her adopter. According to the villagers, transference depends on the child's preference for his real or foster parents. Among the 64 unmarried Wonie villagers in my study period, some 20 are said to have been adopted, but actually only two boys live with their foster parents. Patriliny involves the inheritance of property as well. The most valuable properties in their society, sago stands and coconut stands, are handed down through the male line, usually from a father, alive or dead, to his sons, and, even if the boy is adopted, he will inherit some of his real father's properties.

The traditional age-grade system (Table 5) is still maintained. The system is rather simple, with six and four stages arranged for males and females, respectively. Young children of the *sobijogbuga* stage for boys and *sobijogngamugai* stage for girls are cared for by their parents and other adult villagers.

Table 5. Age-grade system

male	estimated age	female
miid		
	50s	*nanyukonga*
nanyuruga		
	40s	
rugajog		
	20s	*kongajog*
kewalbuga		
	16–17	
yambuga		*ngamugaibuga*
	7–8	
sobijogbuga		*sobijogngamugai*

The boundary between *sobijogbuga* or *sobijogngamugai* and *yambuga* or *ngamugaibuga* depends on the acquisition of verbal communication. Sexual differentiation in behavior tends to be manifest during the *yambuga* and *ngamugaibuga* stage. *Yambuga* boys frequently play with bows and arrows, and, as they become older, sometimes accompany *kewalbuga* males on exploring tours, while the girls tend to spend much time with their mothers and elder sisters who have married, in child care and collecting firewood, and occasionally accompany the married females in fishing, gardening and sago-making. However, neither *yambuga* boys nor *ngamugaibuga* girls make any significant contribution to the subsistence production.

The girls normally marry in their upper teens, when they enter the *kongajog* stage, while boys of around sixteen or seventeen years old advance to *kewalbuga* when the adult men note the maturity of their physical and perhaps psychological development. *Kewalbuga* males leave their parents to live in a separate group, staying at a young men's house called *iit*; during my 1971–72 study period, however, the *iit* houses of Wonie were broken up and the young males lodged separately in vacant houses or in married families' houses other than their parents'. While they learn how to hunt animals, come to know who owns which sago stands and coconut stands, and familiarize themselves with the topography, vegetation, network of paths in their land, and traditional customs, *kewalbuga* depend to a great extent on adult villagers for their daily food. In their twenties *kewalbuga* males usually marry to enter the *rugajog* stage.

Men of *rugajog*, *nanyuruga* and *miid* are heads of households, and women of *kongajog* and *nanyukonga* are housewives, both providing sustenance not only for their own households but also for the community. The boundary between *kongajog* and *nanyukonga* and that between *rugajog*, *nanyuruga* and *miid* are dependent upon physical features such as proportion of gray hair, but formalized social functions do not clearly differ between the stages, while their behavior quite naturally reflects their age.

Demographic Background

One of the basic components of a man-environment rela-

Table 6. *De facto* population of the thirteen Gidra-speaking villages in 1980

village	male	female	total
Rual	48	60	108
Kapal	52	78	130
Iamega	103	134	237
Wipim	79	77	156
Podare	60	74	134
Gamaeve	68	83	151
Wonie	52	55	107
Kuru	60	69	129
Ume	105	70	175
Zim	40	48	88
Peawa	70	67	137
Abam	70	63	133
Dorogori	69	59	128
total	876	937	1,813

tionship in an ecosystem is the volume of human population. The exact figure for the population in the Oriomo Plateau is not clear. The Papua New Guinea Government first carried out a well-planned national population census in 1980, although only one report concerning province-level statistics has been published (National Statistical Office 1981). The estimated population for each linguistic group of the Eastern Trans-Fly Family, which was shown in Wurm's above-mentioned report (1971b), may be suitable for a rough approximation of the population in the region. If so, the Gidra group consisted of 1,600 people in the 1960's. Our 1980 survey disclosed the *de facto* population of the thirteen Gidra-speaking villages as totalling 1,813 (Table 6). Since the area of their land is estimated to be a little less than 4,000 square km, the population density is placed on the level of 0.4 to 0.5 person per square km, with a considerable portion of the land almost unexploited by the people.

When we try the same calculation for the three linguistic groups of the Eastern Trans-Fly Family (Gidra, Bine and Gizra) as a whole, on the basis of Wurm's estimation, the population was 3,750 in a land area of 6,000 square km; thus the population density increased to about 0.6 person per square km. These calculations demonstrate that the population density in the Oriomo Plateau, at least in its inland zone, is lower than one person per square km, a figure which may be more common worldwide among hunter-gatherers than among horticulturalists in the humid tropics. Within Papua New Guinea, this population density is comparable to densities around the tributaries of the Fly, Sepik and Purari Rivers but is considerably lower than those for the majority of the peoples (cf. Ward and Lea 1970; Skeldon 1979).

Along with the population size and density of local groups, human ecological fieldworkers are interested in the population dynamics of the societies studied. For underdeveloped societies, however, whether the size of the population is at a steady state or is rapidly increasing or the degree to which the population has changed has seldom been explored. By means of recently developed simulation analysis for small populations—excellent examples are those for the Yanomamö horticultural people in the Orinoco (MacCluer, Neel and Chagnon 1971) and for the original French population in St. Thomas Island in the Caribbean Sea (Dyke and MacCluer 1973)—life tables of such populations have been estimated. However, even in such demographically-oriented studies, a "stable population" or a model that employed an age composition consistent with the vital rates and that expressed reproductive change on an annual instead of a generational basis (Lotka 1925) was presupposed.

In addition to the absence of a demographic census, the difficulty in estimating past trends in population size for underdeveloped societies derives from the people's unfamiliarity with chronological age. Such problems also existed in studying the Oriomo Papuans. Nonetheless, an attempt was made to identify population trends in the region for the last several decades. For this purpose, I framed a new model of estimation, based on genealogical records, which took into account how many daughters of each woman survived and married. This method is based upon the concept of population replacement through the female line, namely a kind of cohort analysis. For example, in a case dealing with any two generations, for each female of the first generation, her delivery of female children and their deaths, if before childbearing age, are noted. Similar analyses based on the same idea have been made by some anthropological demographers (Crow 1958; Kobayashi 1969; Hassan 1981: 139–142). In the summaries, the balance between the number of females belonging to the first generation who reached reproductive age and that of the second generation is calculated. In other words, this formula identifies the net reproduction rate, which is defined as an intergenerational reproduction rate.

Interviews with the Wonie villagers clarified the genealogical tables back three to five generations. The subjects of this survey were all contemporary Wonie villagers and their ancestors as well as some of their relatives who inhabited adjacent villages. To unify the generation level and to avoid dubious information, the records of the people belonging to three generations, the lowest of which was the present reproductive group, are analyzed. Given the above conditions, females, dead or alive, who had married and had passed reproductive age numbered 88. Judgement as to the completion of reproductive age was based on two criteria. The first was a time of approximately five or more years after delivery of the last child; the second criterion, used especially for sterile and seldom-pregnant women, was to estimate the age of the woman in question through a comparison with those of her siblings and/or other consanguines.

The 88 women are categorized into three groups: Group A = not all offspring have married, Group B = all offspring have married, and Group C = all offspring have completed reproduction. In this categorization, it is natural to assume that the age interval between groups is approximately 20 years. Table 7 shows the number of females belonging to each group, the total number of their male and female children and the number of those who survived to marry. The number of married female children per mother is 1.11 and 1.13 for

Table 7. Total and average number of children of women who have passed reproductive age. The classification of women into three groups is mentioned in the text.

	no.	no. of children			no. of married children[a]		
		male	female	total	male	female	total
Group A	30	40 (1.33)	47 (1.57)	87 (2.90)			
Group B	28	46 (1.64)	32 (1.14)	78 (2.79)	30 (1.07)	31 (1.11)	61 (2.18)
Group C	30	38 (1.27)	35 (1.17)	73 (2.43)	30 (1.00)	34 (1.13)	64 (2.13)

[a] Children who have survived to marry.

Groups B and C, respectively. Assuming that the interval is 20 years, the annual population growth rate can be calculated by the following two methods:

	Formula I	Formula II
Group B:	$e^{20 \cdot r} = \frac{31}{28}$	$(1+r)^{20} = \frac{31}{28}$
Group C:	$e^{20 \cdot r} = \frac{34}{30}$	$(1+r)^{20} = \frac{34}{30}$

Here, e = the base of natural logarithms and r = annual population growth rate.

The values of r were calculated as follows:

	Formula I	Formula II
Group B:	0.005089	0.005102
Group C:	0.006258	0.006278

Thus, we can place the annual population growth rate of this group for the last several decades at 0.5 or 0.6 percent. Accordingly, the population doubling time is from 110 to 140 years. In comparison, the population growth rate for the period when the females of Group A will be completely replaced by their female children may be a little higher, although the final number of deaths before marriage among the offspring of this group is not known.

The annual growth rate of around 0.5 percent should be evaluated from a comparable ecological point of view. Hassan (1973) reviewed and re-assessed various estimates of population growth in nonindustrial or "primitive" societies and concluded that the maximum potential of annual growth is 0.4–1.3 percent. The rate for the Gidra group belongs to the lower margin of this estimate. One may suppose that the 0.5 percent annual rate would result in rapid population growth, but, taking some accidental events like famine and warfare into account, in the long run this level of fecundity may be necessary for the maintenance of population in local groups. Despite Hassan's emphasis on the prevalence of birth control practices among "primitive" groups in relation to the estimated growth rate, there have been no effective practices which are recognized by the Gidra people.

This genealogical-demographic study also clarified the

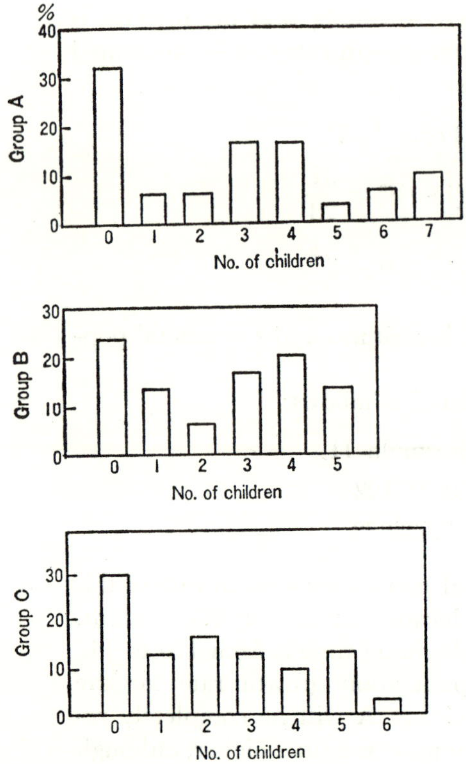

Fig. 3. Percent Distribution of Women according to Number of Married (or Surviving) Children.
The classification of women into three groups, A, B and C, is explained in the text.

number of children delivered by each woman and the exact number of children who survived to marry. The percent distribution of the women by group is illustrated in Figure 3, although, as in Table 7, the children counted for the women of Group A were not those who married but those who had survived up to 1972. The difference in the distribution patterns between Group A and Groups B and C may be attributed to the fact that not all deaths of children for Group A have been recorded. With regard to the influence of the number of marriages per woman on her children, the distribution pattern of women who experienced one marriage and that of women who had two or more marriages differs to a negligible extent, the average number of children for the former being 2.12 and that for the latter 2.17.

The wide variation from a great number of sterile women to a considerable number of women having more than four live births among the Oriomo is of great interest, as is the fact that about 30 percent of the women had no child who survived to marry. In Hassan's above-mentioned (1973) estimation for the annual population growth rate of "primitive" societies in general, 12 percent was given as the proportion of sterile women on the basis of Nag's review work on human fertility in nonindustrial societies (1962: 120–125 and Table 2); strictly speaking, the 12 percent was regarded as an average between low and high sterility, and the records collected in Nag's work showed a great variation in sterility among the societies. Of eight societies in the Pacific area, six were regarded as the high-sterility group, involving over 12 percent of childless or non-pregnant women. For example, the Murut, who are hunters and shifting horticulturalists in northern Borneo, were studied by Polunin and Saunders (1958) for demographic as well as medico-ecological interest. Among the Murut women aged over 40 years, 28 percent had had no live births, a figure which is almost identical to that of Oriomo women. Polunin and Saunders pointed out that three-fourths of the Murut women had been subject to pelvic infection, and as a result the Murut society had been depopulating. Nonetheless, the population of the Oriomo Papuans has increased slightly.

It may be necessary to examine the methodology of this genealogical-demographic study. The most fundamental point is the accuracy of the genealogical records. Since marriage is a most important event in this society, the villagers' recollection of how many of their ancestors survived to marry is believed to be accurate. Putting it in another way, the people tend to forget ancestors who have not married, and this tendency is manifest for females because boys are paid higher regard than girls; this is reflected in the fact that the ratio of "number of married children" per "number of children" for males is lower than that for females, as shown in Table 7. Another problem is the possibility that deceased females who had no child or whose descendants have not survived to the present time have been omitted, since their exclusion would make the estimated population growth rate higher than it is. I paid great attention to this problem, repeating questions to several aged informants. In conclusion, I judge that the estimation is sufficiently valid. The fairly slowly increasing population size involved in this habitat over a long period is useful for understanding the man-environment relationship and adaptive mechanisms on the basis of a cross-sectional study.

Acculturation

Acculturation of the Oriomo Papuans has progressed with the introduction of Christianity, primary school education, a local government system and a cash economy. The following information about these matters is, as a rule, restricted to my study among the Wonie villagers in 1971–72. The Maranatha Mission, with its nearest church in Daru, has had a large following among the villagers since the 1960's, and the missionary patrol visits the village once or twice a year. The villagers themselves occasionally hold Sunday services in the village yard, although not all of them attend. The villagers say that they should not work on Sundays, but actually they frequently carry on their subsistence activities.

One governmental primary school and one mission school of the Seventh-Day Adventists are found on Gidra land, the former in the vicinity of the village Wipim, adjacent to Wonie, and the latter in the village of Peawa along the Oriomo River. These schools were built in the late 1960's. Many boys and girls of Wonie, whose ages range roughly from eight to seventeen, attend the school in Wipim. The school is run by several Papuan teachers from districts (provinces) other than the Western District (Province) in order to educate young people. With the exception of the long-term vacations, pupils live in school dormitories from Monday to Friday and occasionally even on Saturday and Sunday.

In the local government system, each village, or sometimes two or three villages together, is formally represented by a councillor elected by all the adult villagers. He is a member of the Local Government Council of the Oriomo-Bituri area and is in charge of foreign and domestic affairs on the village level. Wonie, together with the village Kuru, has a councillor who lives in Wonie. Actually, except for attending the assemblies held in Daru every two or three months, his particular role was hardly noticeable in performing daily activities or in maintaining community life, since this new system was introduced by order of the Central Government of Papua and New Guinea (the name of the nation before independence in 1975) and has not effectively functioned in the area. Instead, his special role was seen in his intimate contact with the mission and in purchasing goods in Daru and delivering them to the villagers.

Cash has started circulating in the area. The major income is obtained from selling mats and carvings at the market in Daru and from participating in the construction of a wide new road extending from the upper part of the Oriomo River in the east to Morehead on the central upland part of the Oriomo Plateau in the west; this road construction project was undertaken by the Central Government of Papua and New Guinea. The Wonie villagers themselves decided on the days of road-making and worked by themselves. Actually, the work days per person during my study period of seven months totalled ten to twenty days. Due to the sporadic nature of the work, this attempt progressed very slowly and met with considerable setbacks during every wet season. The wage for road-making is paid once a year by the government officers at the working site, although the amount is very small and most of it must be paid back as tax, taxation being one of the major policies which the government has intended to propagate in the area. An adult woman makes one mat a month on the average, and crude carvings are made more infrequently by several young men. Thus, the people's earning is very limited and their purchasing power is still very small. Popular foreign goods are steel axes and knives, clothing such as shirts, shorts, skirts and blouses, kitchen utensils such as pots and bowls, and matches, all of which are commonly used by all villagers. Foreign foods are, however, seldom brought to the village. Thus the daily menu of the people consists of local products.

2. Problems and Research Strategy

Problems

The subsistence and survival of the people of the Oriomo Plateau is of special significance from an ecological point of view. The first characteristic of their subsistence pattern is the heavy dependence upon sago palm (genus *Metroxylon*) which distributes exclusively in the swamp areas from Island Southeast Asia (or the mainland locally) to the Pacific. According to Barrau (1959), species of the *Metroxylon* group appear to have spread in some distant epoch from Indonesia to the Pacific, extending through the Melanesian chain to Fiji, with the exception of New Caledonia. *Metroxylon* is undoubtedly the most important single source of palm stem starch although other genera of palm utilized in a similar manner are available in the tropics worldwide, for instance genus *Caryota* in Asia and genus *Mauritia* in Latin America (Corner 1966). In Melanesia at least 300,000 people rely on the sago palm as their main energy source (Brookfield and Hart 1971: 86), while Lea (1972) estimates that the people dependent upon sago who inhabit Papua New Guinea number 90,000. According to McArthur (1972: 442), all Papua New Guinean people produce 115,000 tons of crude starch annually. Although there is a difference in the number of sago-eaters mentioned in the literature, dependence on sago for subsistence is a distinguishing pattern in the region.

A number of fieldworkers have studied sago-eating peoples inhabiting Southeast Asia and the Pacific, as seen in the review work of Ruddle, Johnson, Townsend and Rees (1978), but there have been few ecological investigations dealing systematically with the interrelationship between sago-eaters and their environment based on a long-term survey in a community dependent on sago at the subsistence level. The most prominent was P. K. Townsend's (1969) study among the Sanio-Hiowe (-Heve) in the Sepik area, northern Papua New Guinea. Based on the limited information, the adaptation of sago-eaters seems to vary from place to place, and thus there is a great need to study systems which have been affected to a lesser degree by acculturation. There are many other related problems which remain to be settled, for example, whether sago exploitation belongs strictly to gathering, or the degree to which it is similar to gathering and/or agriculture as a food procurement strategy, and whether a causal relationship exists between dietary dependence on sago and low population density.

Horticulture plays a lesser role in food procurement for the Oriomo Papuans. This of course accords with the heavy dependence on sago in the region. The people in Oriomo practice slash-and-burn horticulture to raise a variety of crops including banana, taro, yam, sweet potato and cassava. In spite of the existence of crop items similar to other areas in Papua New Guinea, the lesser importance of horticulture in Oriomo contrasts with the predominance of garden products like sweet potato, taro, yam and/or banana as staple foods throughout the region. Hunting with bow and arrow provides the bulk of animal food necessary for subsistence. This also contrasts with the fact that in Papua New Guinea pig raising is popular and hunting is usually recognized as a part-time or leisure activity (R. Bulmer 1969). Thus the combination of such food-getting strategies is unique, not only within Papua New Guinea but also beyond the region. This subsistence pattern seems to involve a low population density. The density of the Oriomo Papuans is less than one person per square km, which is lower than most Papua New Guinean societies which normally have more than ten persons.

The area is of prehistoric significance as well. Man's initial crossing of Wallace's Line from west to east took place in the Upper Pleistocene, and, subsequently, the post-glacial rise in sea level separated Australia from New Guinea (Golson 1971). The most striking difference between the two regions is that in Australia a hunting-gathering subsistence pattern persisted until the European settlement, while in New Guinea horticulture has a long history (White 1971a; Golson 1977). Why this difference arose and continued for so long is still an unsolved mystery. One effective approach has been to investigate how contemporary peoples on both sides of the Torres Strait maintain their subsistence (Walker [ed] 1972; Harris 1977; Jones 1980). Geographically, the Oriomo Plateau is at the front of New Guinea. There are no sago palms in

Australia, but whether there is any causal relationship between sago and horticulture still remains to be determined. The prehistoric significance of sago as a food resource is further stressed by Spencer (1963), who pointed out that the whole of Southeast Asia and the whole of the island archipelago, where rice is predominantly a staple food, were once characterized by a yam–taro–sago cropping complex. Thus, the study of sago palms and their utilization can illuminate one pattern of adaptation in the freshwater zone of the tropics, and this study may also contribute to a greater understanding of the prehistory of Southeast Asia, Melanesia and Australia.

The Papuans inhabiting the Oriomo Plateau have been known to researchers for some time. Ethnographic reports about the people appeared in the early 1900's. Beaver (1920: 60–105) and Archbold and Rand (1940: 159-186) studied the people living along the coast at the eastern and southern fringes of the plateau in a series of travel sketches. Landtman (1927) and Williams (1936) made ethnographic surveys in the Kiwai Islands at the estuary of the Fly River and in the Morehead region several dozen km west of my study region. Their reports, based mainly on indirect information, include brief descriptions of the people in my study area. The major foods and subsistence technologies of the Oriomo Papuans, as noted in their reports, seem to be roughly identical to the present situation, with the dependency on sago as food and sago-making technology being mentioned. Strangely, no systematic ethnographic or anthropological investigation has been carried out in the area since the above reports. The Oriomo Papuans, especially the people of the interior, have never been the subject of intensive, ecological fieldwork.

In addition to the high probability of its providing new information, two local conditions favored my study project. First, the retarded acculturation assured contact with traditional adaptive mechanisms in this ecosystem. Second, each village of 100 to 200 people consisted of approximately 50 to 200 square km, within which the villagers maintained their daily food procurement. Hence, the size of village organization in this local population was neither too large nor too small for the investigation, presupposing that a single researcher (myself) manages to collect various kinds of data and that intensive investigation among the inhabitants of a single village allows a deep comprehension of the holistic subsistence system and adaptive mechanisms.

In addition to the concerns inherent in this local habitat, a methodological concern has been involved in this study. I have decided to apply a new human ecological framework to understanding people's adaptive mechanisms. The central standpoint is in placing a human activity system as a linkage medium between man and his environment. This methodological issue will be touched on in the following section.

Framework for the Man-environment Relationship

The reason why I placed human activity as a key concept in this study should be explained here in a methodological context. Recently developed approaches in human ecology and ecological anthropology have emphasized the qualitative and quantitative understanding of the relation between man and his environment. For measuring and analyzing human life concretely, some human ecologists and/or ecological anthropologists prefer the term "human activity" to the term "human life" *per se* because of its clearer meaning (e.g., Gayden 1974; Watanabe 1977b). In a biological or ecological sense, subsistence activity or the combination of food-getting and food-consuming activities is the most basic component of all activities.

This idea for *the* human ecology parallels animal ecological attainments, the former being hinted at by the latter. In the field of primate ecology, Carpenter's (1934) work on howling monkeys pioneered a detailed understanding of the living system of non-human primates in relation to their habitat. Field investigations of primates along this line, for example, Schaller's (1963) study of mountain gorillas and Altmann and Altmann's (1970) study of baboons, suggested this research design to me. Of special interest to me were their thoroughgoing observations of daily activities.

In the history of quantitative analysis for human activity among social and economic anthropologists, the functionalists have taken the initiative. Richards's (1939) study among the Bemba in northern Rhodesia touched on the time allocated for their horticultural, fishing and hunting labor, while a clear proposal for the systematic, quantitative measurement of subsistence activity was, perhaps, first made in the late 1940's by Digby (1949), who was stimulated by Firth's (1946) fieldwork among the fishermen of the Malay Peninsula, a study which was economically rather than ecologically oriented. Since then, some quantitative approaches to human subsistence activities have appeared in economic anthropological studies, such as Erasmus's (1955) and Tax's (1963) works on Latin American peasant societies.

More elaborate field techniques with a more ecologically oriented framework emerged in the 1950's simultaneous to the development of an ecosystem ecology which focused upon material cycling and energy flow in an ecosystem. One particular technique is called input-output analysis. Its aim is the elucidation of the quantitative balance between the input of energy into subsistence activity and the energy, especially that involved in foods, returned from it, although in fact the input energy is sometimes replaced by the input labor time. The early work along this line involved Carneiro's (1957) study among the Amazonian cassava horticulturalists, Lee's (1968) among the hunter-gatherer Bushmen in the Kalahari, Rappaport's (1968) among the New Guinean highlanders who depend on sweet potato and pig, and Kemp's (1971) among Eskimo hunters. The input-output analysis has flourished as a key concept of the man-environment relationship, while at the same time tending to minimize the role of individuals who differ from each other biologically and behaviorally.

Japanese human ecologists and/or ecological anthropologists including myself (e.g., Watanabe [ed.] 1977a) have paid great attention to the detailed measurement of human activity as a spatiotemporal system structure, studying at an earlier stage fishermen and traditional hunters in Japan in the 1960's and 1970's. Their basic strategy was to replace culture with human activity as the linking medium in the man-environment relationship (Ohtsuka 1970). The work along this line has raised the possibility that the human activity system itself plays a central role in human adaptation, or is regarded as a variable reflecting human adaptation. Their direct contribution was based mainly on results taken from minute observations of the activities of individuals, e.g., rhythm and tempo of activity

(Ohtsuka 1972; Takeda 1972), spatial allotment of activity (Irimoto 1973), individual characteristics in division of labor (Kuchikura 1975) and the transmission of technology and skill through genealogical and neighboring ties (Akimichi 1975). Besides these, I have emphasized individual variations in fishing, diving and hunting activities in terms of labor productivity, which depends on sensory-motor and intellectual skills, the former mainly a function of physical ability and the latter strengthened by work experience (Ohtsuka 1977b, 1979). These findings stressed the functional meaning of individual contributions to human activity systems.

Nonetheless, the study of human adaptation requires an understanding of the people both as individuals and within social groups like households and communities, as defined by cultural anthropologists (e.g., Murdock 1949: 79–90). Human beings do not exist nor maintain their lives as separate individuals but as individuals belonging to and composing multi-leveled structures in social or sociobiological groups. The structure of the adaptive mechanisms which synthesize different levels of organization should be the main target of human ecology, particularly human population ecology as its most distinguished sub-field, one which deals with human population (in biological terms) as a surviving unit (Suzuki 1970, 1977). Watanabe's (1964) study of the Ainu in Hokkaido of northern Japan and Stauder's (1971) study of the Majangir in Ethiopia are good examples of dealing systematically with human groups in relation to food procurement and adaptations although biological and behavioral elucidation of individuals remains to be fully examined. This view appears to parallel Vayda and McCay's (1975) orientation to human ecology and ecological anthropology.

The present study aims at appraising human activity on an individual-to-community basis in a man-environment relationship. A simplified schematic diagram for the major research items and their relationships is illustrated in Figure 4. For this purpose, human activity data were gathered mainly by observation and are analyzed in this volume in relation to such matters as time allocation, input-output balance, space-use, division of labor and/or cooperation, and distribution and consumption of foods. This study also intends to touch on a problem corresponding to the life cycle of man in relation to

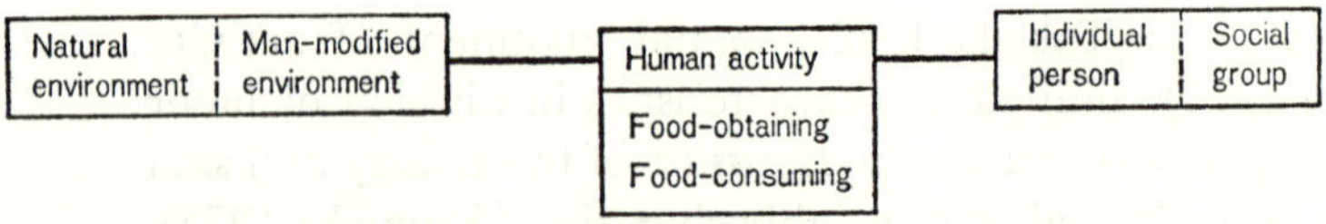

Fig. 4. Schematic Diagram of Major Research Items for Man-Environment Relationship.

human adaptation (Suzuki 1980) although the cross-sectional observations in the present study mainly make possible consideration based on a comparison between individuals of differing age and sex. The above-mentioned framework for human adaptation accords with a new trend in evolutionary studies which emphasizes the importance of functional-ecological systems of contemporary populations (e.g., Peterson and Peterson 1977; Watanabe 1978).

Rapport and Progress of Fieldwork

I first visited the Oriomo Plateau in July 1967, accompanied by R. Hide who then worked for the New Guinea Research Unit of the Australian National University in Port Moresby. This short visit gave me a good opportunity to make up a design for the main fieldwork and to scrutinize the associated methodology. What was established and/or reaffirmed involved two basic components of the field strategy. First, one village was intensively investigated during the dry and wet seasons because, as mentioned previously, each village in the area forms a basic subsistence entity and because the seasonal changes in the environment might affect the people's activity and adaptation. Second, the principal data were collected by direct observation of the people's daily activities, distinguishing each villager from the others.

The main fieldwork commenced in July 1971, and for the beginning six weeks I collaborated with H. Watanabe who carried out a functional-ecological study of bows and arrows (Watanabe 1975). One village called Wonie was chosen as a survey village because of the absence of a primary school and missionary church (several of which existed among villages in the area), its smaller population and the compact spatial arrangement of the dwellings in the village. We feared that such new institutions as schools and churches might directly disturb the traditional life style. The latter two criteria were judged to make the observation study both convenient and profitable.

After the completion of H. Watanabe's study, my own survey continued up until March 1972 with a three-week break from December to January. The first five months fell during the dry season and the latter three months during the wet season. During my fieldwork, I stayed at a lodge used once or twice a year for governmental patrols. It was a relatively easy 50 m away from the main settlement of the Wonie village. Occasionally I stayed out of the village when I went on short trips with the villagers to their working places. The patrol lodge was built by the village people, and its plan and materials were identical to their own houses. In addition to accommodation, the difference in the way of living between the villagers and myself was minimized as far as possible. For example, I usually ate local foods only.

Detailed identifications of the people and their land were necessary for observations on human activity, which was pursued as a key concept in this fieldwork. The beginning several weeks of my survey period were mainly devoted to distinguishing between the 40 married people and more than 60 unmarried people in the village of Wonie, and to exploring the village-land covering 70–80 square km. For the former, I identified age (or approximate age) of the individuals, age-grade, organization of household, marriage experience, genealogical structure, clan system and other sociopolitical organizations, although this work continued beyond the above period. For the latter, I mapped out the village-land on a scale of 1/20,000, including principal paths and some landmarks, as well as the more than 200 localities named by the people. This mapping survey was greatly aided by magnified copies of the aerophotographs made by the Commonwealth Scientific and Industrial Research Organization (CSIRO) of Australia. Many local place names were well known by the people, especially by adult males, and their knowledge greatly helped me in locating and identifying those places.

The details of field-data gathering are described later in association with the results. Here I will outline the methodological framework of the principal research items. The most intensive and time-consuming investigation consisted of surveys on human activity and food consumption. One activity survey was the systematic observation of the people's daily activities, especially food-getting or -producing activities, from morning till evening. This investigation was carried out twice, once in the dry season and once in the wet season, for a period of thirteen successive days each. The other type of investigation of human activity involved following the villagers going to their working places, such as hunting grounds, sago groves and slash-and-burn gardens, and observing concretely what took place there. The food consumption survey for the sampled households was carried out in both the dry and wet seasons for a period of twelve successive days in each season. Besides the above-mentioned investigations, extensively organized investigations covered a wide area, for example, measurement and examination of some physical features and physiological abilities of the people, collection of plant specimens to be identified later by specialists, description and measurement of various implements, and collection of vernacular names.

Methodologically, the field investigation as a whole was characterized by minimizing verbal communication as a tool of gathering information. This approach was effective for two reasons. First, as in animal ecological studies, the intended implications of activities and other matters were omitted, and, second, I had fewer problems since I did not know the local language well. In the region, Pidgin and Motu, which are lingua francas in Papua New Guinea, the former in the northern and central areas and the latter in the southern areas, are not popular. Interview surveys were frequently assisted by local people, specifically an unmarried male of around eighteen or nineteen years old from the village of Wonie who spoke English relatively well and continuously helped me as an interpreter and guide. With a few exceptions, the survey interviews were so designed as to be answered with proper nouns like place names or persons' names.

As mentioned in the Preface, our new group project (Project leader, Professor T. Suzuki) is ongoing and hopes to clarify the adaptability of the whole Gidra group as a population. Actually, a three-month survey was conducted by my colleague and me in 1980, and another six-month survey was also conducted by six researchers, five Japanese and one Papua New Guinean, in 1981–82. The complete findings of this new project will be published later, but a very small portion of the data will be used in this monograph.

2. The Setting

1. Wonie Villagers

Villagers as Biological Beings

One hundred and three people lived in the village of Wonie at the beginning of my fieldwork in July 1971 and 104 people at the end, March 1972. During my stay, there were three out-migrants and one in-migrant: in December 1971 a *ngamugaibuga* female moved to Zim, a northeastern adjacent village, to be married to a Zim *kewalbuga* male; in early 1972 a young couple of *rugajog* and *kongajog* without a child transferred to Gamaeve, a southwestern adjacent village where the wife's parents lived; and in February 1982 a *ngamugaibuga* female migrated from Podare to Wonie to marry a Wonie *kewalbuga*. No deaths among the adult villagers took place during my study period, while three new babies were added to the population. Actually, seven births occurred during the period, but they included three stillbirths and one infant death, three months after birth. Of the three survivors, however, two had passed less than three months of their lives when I left the area in March 1972, and, according to the follow-up study in 1980, one of the three died in infancy.

The reasons for frequent stillbirths and infant deaths remain to be found although they are possibly attributable to bacillary and dyspeptic diseases, nutritional deficiency and hereditary abnormality. A nutritional discussion concerning this problem was begun in another article which analyzed the nutritures of the trace metals in the hair of the Wonie people

Table 8. Villagers by age-grade (November 1971)

male		female	
miid	2	*nanyukonga*	5
nanyuruga	3	*kongajog*	17
rugajog	13	*ngamugaibuga*	19
kewalbuga	9	*sobijogngamugai*	14
yambuga	14	total	55
sobijogbuga	8		
total	49		

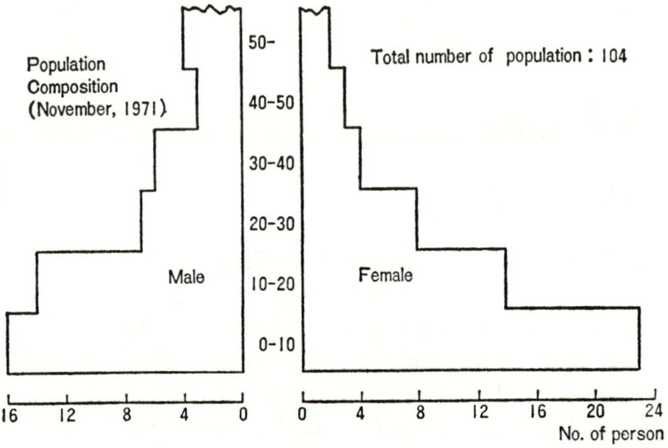

Fig. 5. Age-Sex Structure of Wonie Villagers on the Basis of Their Estimated Age.

(Ohtsuka and Suzuki 1978). The material for this was gathered during this fieldwork, and a summary of the article is given in Appendix 3.

The one hundred and four Wonie villagers at the middle time of the study period, November 1971, are shown by age-grade in Table 8. Despite the people's accurate recollection of birth order among themselves, their chronological age, especially of those past adolescence, is difficult to determine. It may be useful to give the age-sex structure of this population in the usual manner. Estimates of their age based on various criteria such as age-grade, birth order, parent-child and sibling relations, and physical appearance lead to an approximate age-sex population structure (Fig. 5). Roughly speaking, this demographic population pyramid represents a condition of high fertility and high mortality. Considering the unreli-

Table 9. Adult villagers' body height and weight (mean±SD)

	no.	height (cm)	weight (kg)
male	25	166.1±3.35	57.8±5.98
female	11	151.8±3.56	48.3±5.79

Male: *miid, nanyuruga, rugajog, kewalbuga.*
Female: *nanyukonga, kongajog.*

ability of the age estimation, however, the people will be, as a rule, identified not by the estimated age but by age-grade in this volume.

The longevity of the people who passed through the critical time for survival in infancy and did not suffer from contingent deaths like snake-bite is still puzzling. Based on several different kinds of estimations, for example how old the aged person's first daughter's child is and how many years ago the parents of the present inhabitants died, it may be possible to arrive at a rough estimate. As a result, the maximum length of life was supposed to be in the sixties, and the two oldest villagers were examples of this. Nonetheless, the demographic term, life expectancy at age zero, cannot be sought in this study since information concerning loss of life through the life span, especially in infancy, was scarce and unreliable.

The body height and weight of the people were measured. As shown in Table 9, the measurements of the adults, i.e., married men and *kewalbuga* for males and married women for females, average 166.1 cm in height and 57.8 kg in weight for males, and 151.8 cm and 48.3 kg for females. According to data for eleven Papua New Guinean peoples among whom an anthropometric study was undertaken (Malcolm 1970; Walsh 1972; Littlewood 1972), the average adult male measurements ranged from 148.8 to 162.5 cm in height and from 43.4 to 60.5 kg in weight. It is not possible to strictly compare these figures with various people's ages, but when we do make comparisons the Oriomo Papuans are somewhat taller, whereas their weight is within the range mentioned. The Rohrer index (weight/height^3) giving an indication of slimness is 1.26 (as kg/cm^3 $\times$ 10^5) for the Oriomo; this value is at the lower margin of the eleven-group range between 1.23 and 1.62. As for the children, eleven boys of the *yambuga* stage were studied, and their ages were determined based on one-year intervals

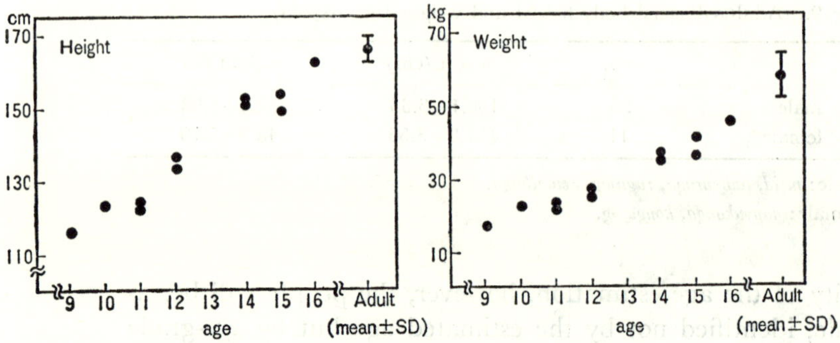

Fig. 6. Body Height and Weight of Male Villagers of Wonie.

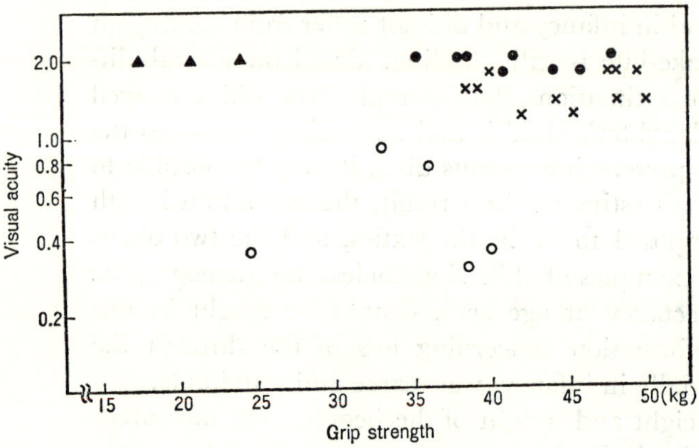

Fig. 7. Visual Acuity and Grip Strength of Male Villagers of Wonie.
▲, *yambuga*; ●, *kewalbuga*; ×, *rugajog*; ○, *miid* and *nanyuruga*.

with some reservations for accuracy. Bodily growth, from a cross-sectional data set (Fig. 6), shows close proximity to the adult level at the time of entrance to *kewalbuga* grade, around 16–17 years old, with height preceding weight.

The measured physiological capabilities were grip strength, recognized as a relatively steady indicator of muscle strength, and visual acuity as an indicator of sensory ability. The measurements for males are shown in Figure 7, in which each individual is indicated by a different mark corresponding to his categorized age group, *miid* and *nanyuruga en bloc*, *rugajog*, *kewalbuga*, or *yambuga* (the oldest three individuals only). Inter-group comparison demonstrates that *rugajog* and *kewalbuga* surpass the other two groups, while *rugajog* is superior to a small degree to *kewalbuga* in muscle strength and inferior in visual acuity. The *miid* and *nanyuruga* group lags behind in both measurements, whereas *yambuga* is far inferior in muscle strength in spite of the superiority of visual acuity. In summary, the *yambuga* boys who are socially recognized as "children" are clearly inferior to adult men in physical strength; the *kewalbuga* males almost all attain a full growth comparable to the young married men; after the peak at *kewalbuga* and *rugajog* stages, the physical strength and sensibility become weak during *nanyuruga* and *miid*. Female physiological features appear to follow a pattern similar to males in relation to aging, although not many individuals were studied.

Villagers as Social Beings

In addition to recognition as a member of the Wonie village, each villager is socially labeled by his or her clan, age-grade and status in a household. Affinal relation is also added in the case of the married people. As shown in Table 10, the number of clans to which the Wonie villagers belong totals eleven although ninety percent of them are members of four dominant clans, Dorem, Wammopam, Dantungam and Gameam. Regarding the spatial arrangement of clan-lands in the Wonie village-land, the Dorem clan-land is completely located within it, and five other clan-lands occupy peripheral

Table 10. Wonie villagers by clan (November 1971)

clan	married		unmarried		total
	male	female	male	female	
Dorem moiety					
Dorem	5	2	16	10	33
Wammopam	1	6	5	8	20
Kameam	1	1	1	0	3
Rekam	1	0	0	0	1
Damuram	0	1[a]	0	0	1
Midam	0	1	0	0	1
Morob	0	1	0	0	1
Wipmam	0	1	0	0	1
Dantungam moiety					
Dantungam	5	3	5	11	24
Gameam	4	5	4	4	17
Rayaiena	1	1	0	0	2

[a] Widow

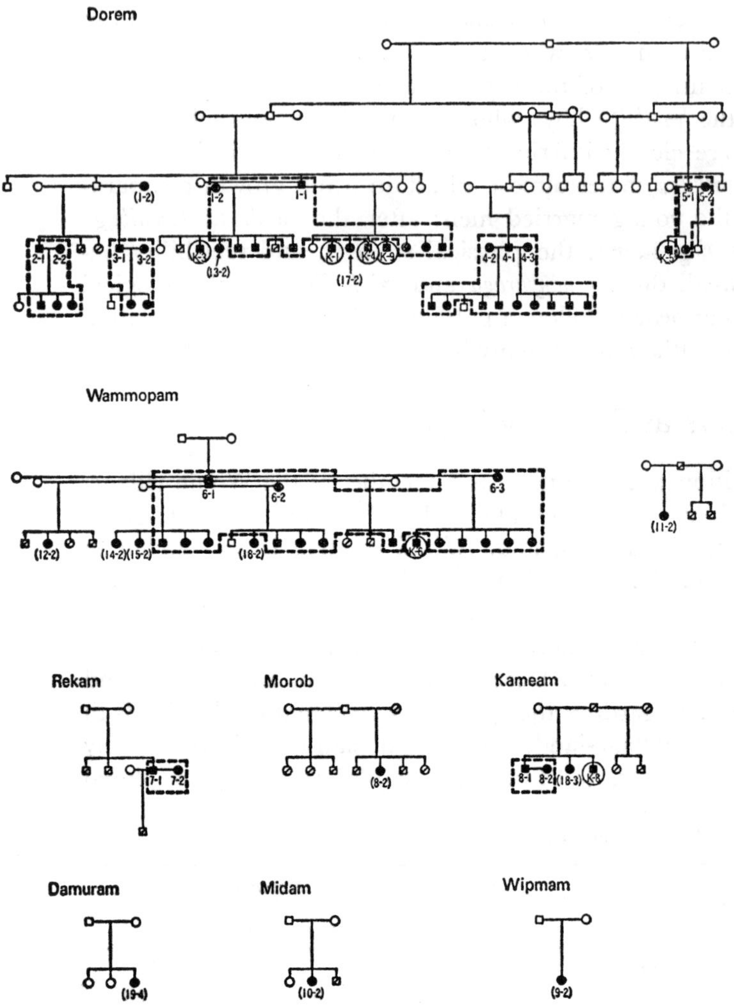

Fig. 8. Genealogical Charts of Wonie Villagers.
Clan names are shown above each chart. The Dorem moiety is on the left side and the Dantungam moiety on the right side. In the charts, spouses and children of the spouses whose descendants did not or do not inhabit the Wonie village are excluded. The numerals show the contemporary adult villagers of Wonie, and the numerals in parentheses show the women who have married (or remarried) and belong to other genealogical charts.

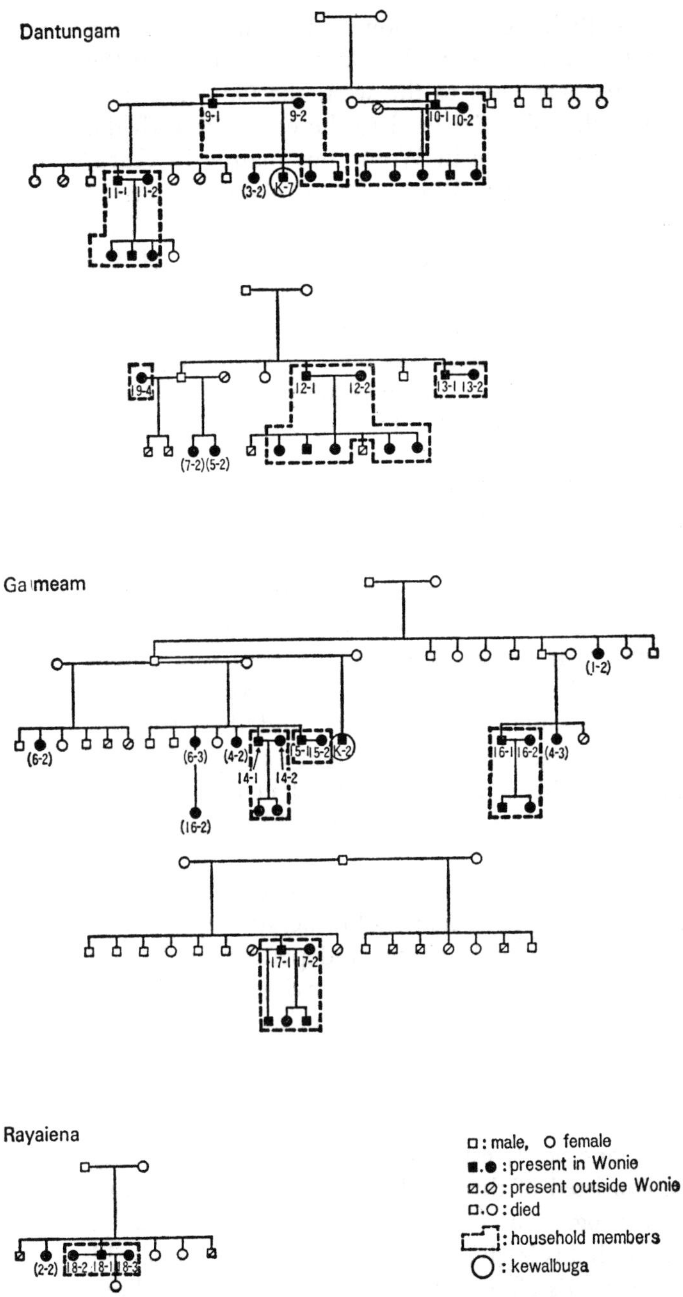

Dantungam
9-1
9-2
10-1
10-2
11-1
11-2
(3-2)
K-7
19-4
12-1
12-2
13-1
13-2
(7-2)(5-2)
Gaimeam
(1-2)
(6-2)
(6-3)
(4-2)
14-1
14-2
15-1
15-2
K-2
16-1
16-2
(4-3)
(16-2)
17-1
17-2
Rayaiena
(2-2)
18-2
18-1
18-3
□ : male, ○ female
■.● : present in Wonie
⊠.⊘ : present outside Wonie
□.○ : died
: household members
◯ : kewalbuga

areas of it while their major portions extend out of it: the Kameam land to the northeast, the Dantungam land to the north, the Wammopam land to the west, the Gameam land to the south, and the Worodai land to the east, although the Worodai people do not live in Wonie but mainly in Kuru, an eastern village adjacent to Wonie. Of the eighteen married males (Table 10), sixteen belong to the above six clans. Two exceptions are a man of the Rekam and another of the Rayaiena. The former was born in the village Zim, where his clan-land is located, and transferred to Wonie at his marriage with a girl of Wonie. The latter was born in Wonie when his parents had transferred there and was adopted by a Dorem man of Wonie, the Rayaiena land being located far northwest of Wonie.

The basic unit of daily living, involved with production and consumption, is a household consisting of a husband and his wife or wives and their unmarried children. *Kewalbuga* males leave their parents to live in a separate group and, except for hunting game, depend on married villagers for their food, especially staple plant foods. Actually, the *kewalbuga* obtain food from any married villager, the majority from male and female relatives, consanguineal or affinal, and their spouses. They are thus recognized to a great extent as dependants of the village group in this sense, despite the fact that they are recognized by the villagers not as children like *yambuga* but as "full-grown-to-be."

Figure 8 shows all villagers on short-cut genealogical charts, with members of each household enclosed by a dotted line and each *kewalbuga* male circled independently. In this figure, each individual who has been married is labeled by a household number plus a number showing status in the household, i.e., 1 = household head, 2 or 3 = housewife and 4 = widow. *Kewalbuga* males are labeled by K plus a number of decreasing order of age among them. The nineteen households, excluding one newly established at the later stage of my study period, consisted of three polygynous families with two wives, fifteen monogamous families and one widow's family. A list of the households is given in Table 11, where household composition, member's clan and the location of the dwelling house are given. The household numbers are the same as those

Table 11. A list of households (November 1971)

Household No.	no. of members	head		housewives				no. of children[c]				location of house
		age-grade[a]	clan[b]	age-grade[a]	clan[b]	age-grade[a]	clan[b]	YB[a]	SB[a]	NB[a]	SN[a]	
1	8	M	Dorem (D)	NK	Gameam (T)			3	1	2	0	Ginar
2	5	RJ	Dorem (D)	KJ	Rayaiena (T)			1	0	2	0	Ginar
3	4	RJ	Dorem (D)	KJ	Dantungam (T)			0	0	0	2	Ginar
4	12	RJ	Dorem (D)	KJ	Gameam (T)	KJ	Gameam (T)	2	4	3	0	Ginar
5	3	NR	Dorem (D)	KJ	Dantungam (T)			0	0	1	0	both
6	15	M	Wammopam (D)	NK	Gameam (T)	NK	Gameam (T)	4	0	5	3	Ginar
7	2	RJ	Rekam (D)	KJ	Dantungam (T)			0	0	0	0	Ginar
8	2	RJ	Kameam (D)	KJ	Morob (D)			0	0	0	0	Ginar
9	4	NR	Dantungam (T)	NK	Wipmam (D)			1	0	1	0	Gimurugi
10	7	NR	Dantungam (T)	KJ	Midam (D)			0	1	3	1	Gimurugi
11	5	RJ	Dantungam (T)	KJ	Wammopam (D)			1	0	0	2	both
12	7	RJ	Dantungam (T)	KJ	Wammopam (D)			1	0	2	2	Ginar
13	2	RJ	Dantungam (T)	KJ	Dorem (D)			0	0	0	0	Ginar
14	4	RJ	Gameam (T)	KJ	Wammopam (D)			0	0	0	2	Ginar
15	2	RJ	Gameam (T)	KJ	Wammopam (D)			0	0	0	0	Ginar
16	4	RJ	Gameam (T)	KJ	Wammopam (D)			0	1	0	1	Ginar
17	5	RJ	Gameam (T)	KJ	Dorem (D)			1	1	0	1	Ginar
18	3	RJ	Rayainea (T)	KJ	Wammopam (D)	KJ	Kameam (D)	0	0	0	0	Ginar
19	1			NK	Damuram (D)			0	0	0	0	Ginar

[a] Abbreviations for age-grades: M=*miid*, NR=*nanyuruga*, RG=*rugajog*, NK=*nanyukonga*, KJ=*kongajog*, YB=*yambuga*, SB=*sobijogbuga*, NB=*ngamugaibuga*, SN=*sobijogngamugai*.

[b] (D)=Dorem moiety, (T)=Dantungam moiety.

[c] Children's clan is their father's.

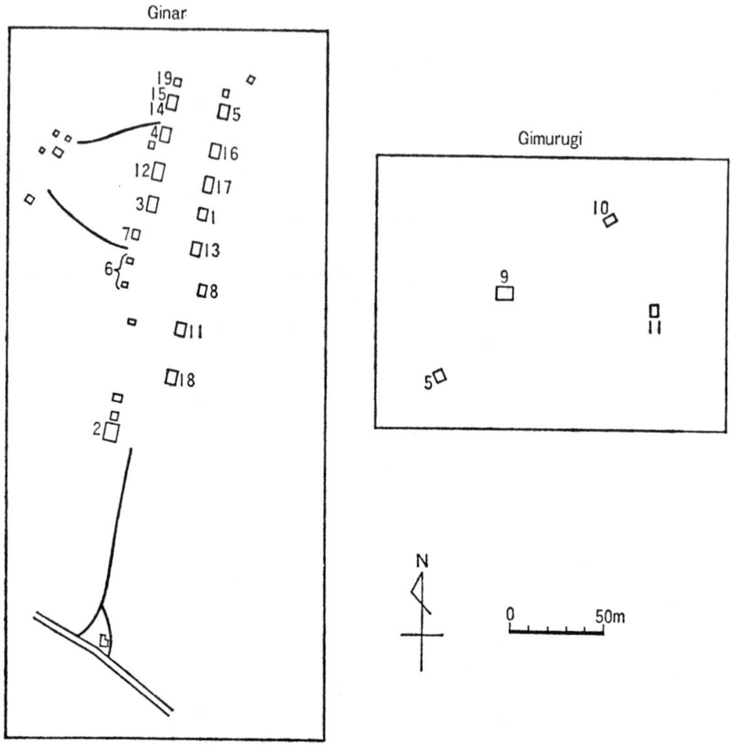

Fig. 9. Village Settlements at Ginar and Gimurugi, Showing Dwelling Houses. Five rectangular marks in upper left at the Ginar settlement indicate broken *iit* (dwelling house of *kewalbuga* males), and the L-shaped mark at the bottom at Ginar indicates a hut for governmental patrols, which was occupied by me.

in Figure 8 and Figure 9 and will be used consistently in this volume.

The members of these households consider themselves a single village group although the village has two permanent settlements, the large one at Ginar and the small one at Gimurugi, five km north-north-east. Gimurugi was the main settlement until more than ten years ago, when the village settlement was moved to Liig close to Gimurugi, as will later be explained in detail. After that, the settlement at Liig was abandoned in 1967, when the majority of villagers moved to the present main settlement at Ginar, while two households (Nos. 9 and 10) separated to return to Gimurugi, and two other households (Nos. 5 and 11) built two houses each at Ginar and at Gimurugi. The former two households, whose heads are real brothers, do not have a house at Ginar, and thus their daily activities are centered at Gimurugi. Members of the latter two households divide their time between Ginar and Gimurugi. The spatial arrangements of the dwelling houses in the two settlements are illustrated in Figure 9. No special rules concerning the kin- or affinity-relation or the clan system seem to apply to the arrangements of the houses.

At the village level, no traditional chieftainship has existed, and thus the villagers are not governed by special persons or organizations. In other words, the spectrum of social organization from individual to village is very simple; aside from the household unit, no social group of individuals has a formal structure or functions together, nor is there a group of households except for the village itself. Naturally, however, kin- or affinity-bonded individuals and clan members tend to maintain close relations and cooperate. In addition, unmarried villagers of the same sex and similar age tend to spend much of their time together, and *kewalbuga* males especially make groups frequently for play and for many other activities.

A profile of the households, in which special attention is paid to the life history relevant to birth place and inter-village migration, is given as follows (cf. genealogical trees in Fig. 8).

Household No. 1. The head of this household is a *miid* and the second oldest among the villagers. He is *mopruga* of the Dorem clan. He was born in the Wonie village-land and has never experienced a lengthy stay out of Wonie. All of his three wives, two of whom have already died, were born in villages other than Wonie, the first wife in Kapal, the second in Kuru and the third in Gamaeve. One of his sons, who was the oldest *kewalbuga* (Indiv. No. K-1), married a girl from Podare in the late stage of my fieldwork.

Household Nos. 2 and 3. The two household heads, who belong to the Dorem clan, are brothers by different mothers and were born in Wonie. The wife of Household No. 2 was born in Wonie, while the wife of the other household was born in Iamega and is a daughter of Indiv. No. 9–1, who is a present member of Wonie.

Household No. 4. The household head, a Dorem man, and two of his wives were born in Wonie. Both of the wives belong to Gameam, one of the predominant clans in Wonie.

Household No. 5. The household head is also a Dorem man and was born in Wonie although his genealogical relationship is far from the other four Dorem household heads. Two of his wives, one of whom has already died, belong to the Dantungam clan and were born in Iamega, a northern adjacent village. This household possesses two dwellings, at Ginar and at Gimurugi.

Household No. 6. This household head, the oldest among the contemporary villagers, was born in Podare and transferred eastwards to Wonie after his first marriage. Nonetheless, his arrival in Wonie took place more than twenty years ago, and his efforts at assimilation into the Wonie village have progressed well; this is typically seen in the expansion of his sago and coconut groves there. His clan is Wammopam, and he is a *mopruga.* The Wammopam are said to have nearly disappeared in his adolescence and thus, with thanks to the assistance of the people of the same moiety as Wammopam, he got married to six women by means of women exchange in order to have a number of followers. Of the six wives, four were born in Podare or Wipim, close to Podare, while the other two were of Wonie origin, both belonging to the Gameam clan.

Household No. 7. The head of this household belongs to the Rekam clan, whose members aggregate in the village Zim, where he was born. After the death of his first wife who was also born in Zim, he transferred to Wonie to marry a Wonie girl. Although he had been in Wonie for several years, this was not enough time for his own sago and coconut stands to mature. Thus, he and his wife sometimes have to visit the Zim village-land to harvest his sago and coconut.

Household No. 8. This was the last household that was established, without taking into account a new marriage which took place during my stay. The household head was born and grew up in Wonie, but his father inhabited Iamega village except for short-term stays in Wonie. His wife came from Gamaeve, to which this married couple transferred in the later stage of my study period.

Household Nos. 9, 10 and 11. All of the three household heads belong to the Dantungam clan, most members of which live in Iamega. The heads of Household Nos. 9 and 10 are brothers in the *nanyuruga* stage. Both of them were born in Iamega village-land, although at that time the Iamega main settlement stood closer to Gimurugi than it now does. These two brothers joined the Wonie people during their *rugajog* stage about a dozen years ago when the Wonie village stood at Gimurugi. They followed the other Wonie people at time the village moved from Gimurugi to Liig. When the Wonie villagers built their present settlement at Ginar in 1967, however, the brothers left the majority to return to Gimurugi. The head of Household No. 11 is a son of the elder brother in question (Indiv. No. 9-1) and was born in Iamega and transferred to Wonie with his parents during his boyhood. Household No. 11 possesses two houses, one at Ginar and one at Gimurugi. The birth villages of the wives of the three households were as follows: one in Wonie and one in Iamega for Household No. 9; two in Wonie and one in Iamega for Household No. 10; and one in Kuru for Household No. 11.

Household Nos. 12 and 13. The two household heads of the Dantungam clan are brothers, and their births took place in Wonie land. They are unable to trace back the kin-relation between their father and the father of the two heads of Household Nos. 9 and 10, and thus their feeling of intimacy is of less special strength. The wife of the elder brother is a daughter of a *miid* man, Indiv. No. 6-1; she was born in Podare land but lived together with her parents in Wonie for a long period before marriage. The wife of the younger brother is of Wonie origin and is a daughter of Indiv. No. 1-1.

Household Nos. 14, 15 and 16. Of the three heads, all of whom belong to the Gameam clan, Indiv. Nos. 14-1 and 15-1 are brothers of the same parents while Indiv. No. 16-1 is a cousin of theirs. All of them were born in Wonie, and all of their spouses were also born there and belong to the Wammopam clan, although the wife of Household No. 16 is an illegitimate daughter of Indiv. No. 6-3 and was looked after by the mother's husband (Indiv. No. 6-1), who is also the father of wives of Household Nos. 14 and 15.

Household No. 17. The household head belonging to the Gameam clan was born in Gamaeve and transferred to Wonie at the time of marriage. His wife is a daughter of Indiv. No. 1-1 and was born in Wonie. While only several years had passed after his transference, his sago and coconut groves, including those which were inherited from his father, are distributed in the southern part of the Wonie village-land, close to Gamaeve village-land. Accordingly, his sago and coconut harvests can, to a considerable degree, be performed within Wonie land, with additional harvesting trips to Gamaeve land.

Household No. 18. The husband and two of the wives in this household were born in Wonie. In comparison with the other two polygynous families, the head of this household spends much more time with one of the wives (Indiv. No. 18-2). Consequently the other wife (Indiv. No. 18-3) carries on production activities fairly independently or with villagers other than her husband.

Household No. 19. This household involves only a widow, who was born in Kapal and transferred to Iamega, where her husband was born and died, at marriage. Her transference to Wonie may have been affected by the fact that her husband's brothers (Indiv. Nos. 12-1 and 13-1) lived there although she manages to make a livelihood without special dependence on them. She has two sons, both of whom are *kewalbuga* and who remain in Iamega.

From Cradle to Grave

When a woman becomes pregnant, her behavior seems to continue unchanged virtually until childbirth. For instance, I observed a woman deliver a child on the trip to a sago grove; she suddenly left the group to crouch down on a spot several m aside from the path. Two women in the party attended her, and later several women were summoned from the village, while the husband was stationed around the spot without any specific duties. In this case, the woman's child was stillborn on the same day. One other case, which I observed sporadically, resulted in a live birth. The pregnant woman left the village alone in the evening for a spot some tens of meters away, where she made a simple shelter. The delivery spot is locally called *munapu*. The place of *munapu* in Wonie is not specifically fixed although most deliveries are actually made in a nearby area as long as the woman stays in the village until just before delivery. It was noted that the woman in question had performed all productive and domestic activities as usual until leaving for the delivery spot. From nightfall of the day in question, a number of women, including *ngamugaibuga* girls, came and went in twos or threes between the village and the delivery spot, and several women spent the night there. On the evening of the second day, she gave birth. The manner of giving birth in the latter case seems to be more common.

Mothers whose children are not stillborn stay a couple of days at the delivery spot before returning to the village. Several days after the birth, the mother will begin performing light work such as fetching firewood and harvesting a small amount of food from the nearby gardens, and her full participation in any food-getting activities will commence less than a month thereafter. For the beginning three or four months the baby spends much time in the mother's arms or in a special basket called a *suta* made of pandan leaves. The incidence of infant death tends to be concentrated in the first several months. Corresponding to this, when a baby has survived for one-fourth or half a year he or she is named at a feast held in the parents' house. A number of villagers attend the feast, and one of them, usually an adult male other than the baby's father, gives the baby a name. As far as I know, the selection of the godfather is not based on any specific rules. The given name is usually borrowed from the godfather's ancestors or the baby's ancestors. After the naming, the baby is recognized by the villagers as *sobijogbuga* for a boy or *sobijogngamugai* for a girl.

The periods of lactation and continual care of the baby last for about two years, but after a few months the baby's food items broaden and other villagers assist in child care. When lactation is over, the child begins to move around nimbly in the village yard or in the place where the mother works. The child may sometimes be left in the village during a mother's short stay outside the village. At approximately four years of age, a boy is provided by his father with a bow made of tree branch and arrows made of grass stem to practice shooting with his playmates of a similar age. According to my observations, no specific playthings are found among the girls in this age group. From this age, same-sex play groups become more common.

The age of seven or eight is the time of age-grade change from *sobijogbuga* to *yambuga* or from *sobijogngamugai* to *ngamugaibuga*. Concurrently, the child begins attending primary school, where the full period of education is six years, although in fact some of the school-age children completely refuse schooling and others attend infrequently. In this stage, children's behavior depends largely on their parents. They usually follow the parents in long-term work trips to sago groves, coconut groves and gardens. On the other hand, their play sphere tends to extend to the surroundings of the village-yard and beyond the parents' sight. In later stages of behavioral development boys typically follow *kewalbuga* males on expedition tours and girls assist their mothers or elder sisters in tasks such as fetching firewood, fetching drinking water, child care, cooking, and even in production work. Nonetheless, their direct contribution to food production is still negligible, and, conversely, the adults never assign participation in labor to unmarried children.

With the recognition of his physical and perhaps psychological development by the married men of the village, a *yambuga* boy of sixteen or seventeen years enters the *kewalbuga* stage. When a boy becomes a *kewalbuga*, he is given a *kewalbuga* name, by which he will be called later on. The name giver, who will be related to the *kewalbuga* as his *omugam*, is usually a *rugajog* man belonging to a clan other than the *kewalbuga*'s, and the name given is normally the giver's ancestor's or the *kewalbuga*'s ancestor's. While it is said that the *omugam* is obliged to look after and train the *kewalbuga*, his role is not obvious, at least at the present time. Although no special feast is prepared for a boy's entrance to *kewalbuga*, the married people expect the *kewalbuga* male to be a "full-grown" man in the future, that is, to manage a household and contribute to overall village life. Several changes in his behavior take place concurrently. First, he leaves his parents' house to live in groups with his age-mates. Second, he becomes an active hunter, and his hunting catch will be presented to all the villagers. Third, the bulk of his daily plant foods is provided not only by his parents, but also by his *omugam* and any other married people. Last, he is subject to food tabu; this is explained later. I observed one example in Indiv. No. K-9, who became a *kewalbuga* in September 1971.

Traditional marriage custom is based on exchange of women between two bridegrooms who belong to different moieties. However, I observed one independently organized marriage, in which a *kewalbuga* (Indiv. No. K-1) of Wonie brought a *ngamugaibuga* from the Podare village with no clear beforehand signs. Without any kind of wedding ceremony, their marriage seemed to be promptly accepted by all Wonie people, and the bride never visited her birth village during the remaining few months that I was in the region. According to the husband's statement during the 1980 follow-up study period, he had obtained agreement to the marriage from his father who was by chance a *mopruga* of the Dorem clan which he belongs to, although no negotiation with the wife's parents had been made. For a couple of months after the marriage, their daily food came mainly from their fetching of crops from the husband's relatives' gardens and their participating in sago-making parties involving his relatives. As a traditional custom, the bridegroom has a special relation with his bride's father, the former thinking much of the latter and sometimes helping him in presenting foods, assisting in garden-making, etc. At the same time, the bridegroom should not call on his wife's mother nor take a position close to her on any occasion. These relations with the wife's parents are carefully observed.

My knowledge of pregnancy among unmarried females, divorce, polygynous marriage and remarriage is limited, although information from several different sources may make clear some aspects of these matters in the society. The prohibition against sexual intercourse other than between a married couple applies to married and unmarried women; it is severe for married females but not so severe for the unmarried. When an unmarried *ngamugaibuga* becomes pregnant, she may normally marry the partner, married or unmarried, before the child's delivery; alternatively, she may care for the baby herself until she gets married and brings the baby into the new household; another possibility is that she gives birth without marriage, and then her child is adopted and fostered by a married couple after the lactation period. Whatever the case, the child is recognized by the people as a member of their society without any distinction from children of normal married couples.

Divorce occurs infrequently, and the predominant reasons triggering it are a husband's discontent with his wife's failure to produce children and an unfamiliar feeling and/or jealousy among the wives of a polygynous family. One of the active reasons for having plural wives is said to be the failure of the wife or wives to produce children although some other matters, like the husband intending to increase the number of his clan members, or the husband having a number of sisters to be exchanged, seem to be among the major reasons. Males can remarry soon even when they lose their wife in *nanyuruga* stage, while females in *kongajog* stage, or roughly of reproductive age, normally remarry, but those in *nanyukonga* stage may remain widows.

Death, together with childbirth (more exactly survival of the infant beyond the critical period of the beginning several months), is recognized by the people as a matter of utmost concern. Stillbirth does not cause special sensation in their society. Based on my observations, the death of an infant who had not yet been named brought about a mournful situation in the village, but what occurred involved only the mother and several villagers crying over the death for a day and burying the body under the ground several hundreds of meters apart from the village. Although no adult villager of Wonie died during my stay, the people said that the more aged the dead, the greater the mourning feeling and the larger the funeral held. In the case of the death of a *kewalbuga* in Wipim village, a number of Wonie people made a call of condolence; they said that all in attendance, involving all Wipim people and many others from adjacent villages, cried for several days.

Since there is no cemetery in the area, burial places are selected casually. The people rarely call at a grave, but its location remains in their memory for long time. The people remember well ancestors two or three generations back, especially male ancestors belonging to their own clan, for their names and behavioral characteristics, which are known either directly or by tradition.

2. Village-land and Activity Space

Village-land

The village of Wonie is located in the center of the Gidra land, surrounded by six Gidra-speaking villages, i.e., Iamega,

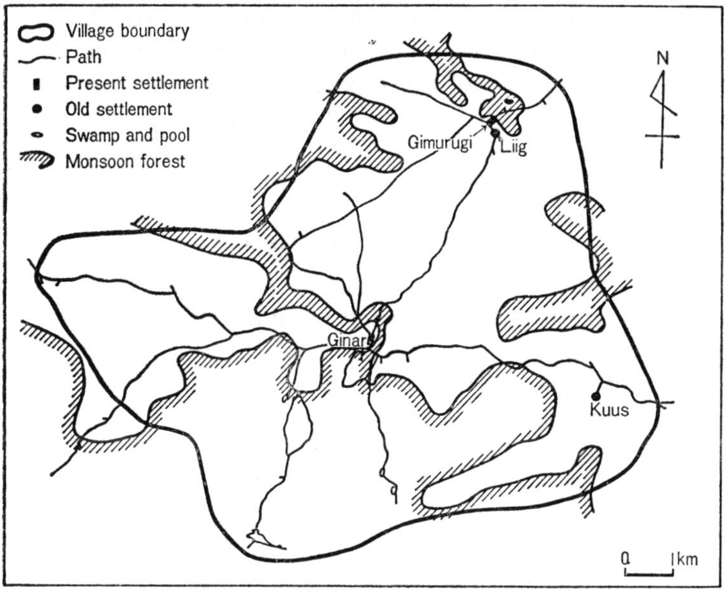

Fig. 10. The Village-land of Wonie.

Wipim, Podare, Gamaeve, Kuru and Zim, which are a half-day trip away (cf. Fig. 1). A space of about 70–80 square km is recognized as the village-land of Wonie, as is shown in Figure 10. As far as I observed, however, the zones close to the boundaries of village-lands were seldom used by the people from either village. Geomorphologically the Wonie village-land is fairly flat, the complicated network of creeks making it slightly uneven. Two-thirds of the northern part is on the Morehead Ridge and the remainder slopes gradually towards the coastal plain in the south.

The two predominant types of vegetation are woodland (*bua*) in low altitudes and savanna (*yap*) in high altitudes; roughly speaking, the southern part of the village-land is comprised of the former, and the central and northern part, the latter. Though not indicated on the map, the land has many small creeks as well as patched woodlands (*kuur* and *buayap*) and small riverine forests (*bojogbua*) in the savanna. Several waterpools, some hundreds of square m in area, occur in the woodland, and most of them are covered with sago groves. The paths indicated on the map are the only distinct links with other villages and/or most of the large sago groves.

Slash-and-burn gardens are usually made along the paths, and, alternatively, when a new garden is cleared in a place remote from the paths, a small track extending to the nearest point on the path gradually becomes distinguishable owing to occasional travel. There are a number of inexplicit paths, some of which lead to old village settlements and were once frequently used. Some others lead to sago and coconut groves, which are infrequently visited. For trips to hunt, fish, or gather food and other useful materials, the people prefer to use these paths as far as possible. The creeks in the village-land are not so wide as to require canoes for crossing, unlike those of some adjacent village-lands. Instead, several simple log bridges span larger creeks.

The people are very well acquainted with their own village-land. This is reflected in their ability to minutely distinguish more than 200 place names within the village-land and to be aware of the ever-changing conditions of the environment such as the amount of water in the creeks, the degree of dryness of the grasses in the savanna and the times when fruit (some of which is eaten by game animals, especially cassowaries) is in season. They are also able to trace remote paths, which are completely indistinguishable to outsiders.

Throughout the village-land, no area is exclusively owned or used by specific persons or specific groups. The only exception is a small area of less than one square km called Goara, which the people consider sacred based on their mythology. Goara, which is located three km south of the present village settlement of Ginar, is a depression with two channels extending from the rim several m above the bottom towards the surrounding lowland, and it thus forms a waterpool in the wet season and a barren land with only short grasses in the dry season. All villagers except married males and *kewalbuga* males are strictly forbidden to visit there. However, Goara is of no specific economic importance and provides the people with fish for a relatively short period, i.e., two or three days in the early dry season, when a large number of fish congregate in the decreasing water of the Goara depression. The fishing procedure at Goara for the 1971 dry season is described in Appendix 4.

The village people freely exploit any place for its natural resources, either food or other useful materials, by means of hunting, fishing or gathering. In contrast, conditions differ for horticulture and sago and coconut exploitation, since it is natural that the areas for these food-getting activities are owned by an individual or a household and more or less managed by the owner. Furthermore, there are remarkable differences in use of land between horticulture and sago and coconut. This appears to result from the fact that it takes less than one year for most garden crops to mature, but ten to fifteen years for sago and coconut. All garden plots are moved frequently, as with most slash-and-burn horticultures in the world. On the other hand, sago and coconut grow naturally or are planted in semi-permanent groves. Whoever clears a garden plot and plants cuttings and shoots owns the garden and the crops; usually garden crops are owned by a husband and his wife or wives. Gardens are not inherited. In contrast, sago and coconut stands are individually owned by married men. The ownership of naturally propagated stands is held by the owner of the mother palms. Sometimes sago and coconut are inherited. When a man marries he inherits some sago and coconut stands from his father, alive or dead. These vary in number according to the number of stands owned by his father and the number of his brothers. After this, he will increase the number of his sago and coconut palms by planting shoots and seedlings. In terms of spatial arrangement, gardens are usually located within half an hour's walk from the village, whereas sago and coconut groves lie scattered over the Wonie land, with no regard to distance from the village.

Forest in savanna and along rivers (*kuur* and *bojogbua*, respectively) and also relatively densely-wooded savanna (*yap*) are selected for slash-and-burn gardens. Areas suitable for gardens seem very vast around the village, even if every household clears one or two new gardens each year. The location of the garden is decided by the head of the household that owns it. Sago groves grow exclusively on swampy land, and there are several such large areas in Wonie territory. In most of these, sago palms thrive, and the bulk of Wonie sago comes from these large groves, although there are many small natural groves as well. Sago palms give off a number of shoots from their root system after bearing fruit, and therefore sago stands tend to be naturally dense. The people can make small groves along the creeks by transplanting sago shoots, but these attempts have met with only partial success, perhaps because of the lack of any large permanently-running river in Wonie land. Coconuts, which are suited to dry land, grow in various vegetations. While coconut palms propagate naturally from fallen fruit, the people usually transplant the seedlings as they prefer. The possibility of expanding the area of coconut groves is very great, although in fact only a small part of such land has been exploited, in accordance with the required amount of fruit.

The village settlement in the area has moved about every ten years. The present as well as the old village settlements stand on the Morehead Ridge, and the areas for the settlements are a little higher than their surroundings in the micro-environmental setting. The present village at Ginar was established in 1967 (Photo 2). For the 15–20 years prior to that, the people changed the village site three times. The oldest site was Kusu around five km east of Ginar, the next was Gimurugi around five km north-north-east of Ginar, and the last was Liig very close to Gimurugi. According to the villagers, the transfer from Gimurugi to Liig occurred unexpectedly because of several unfavorable accidents including some deaths, while the other two moves were mainly made by popular consent and were due to the deterioration of the dwellings. In my view, however, the following reason should be added to their own conscious reasons for moving the village: overexploitation of the environment close to the village in terms of horticulture, hunting and gathering. As mentioned previously, two households (Nos. 9 and 10) did not move from Liig to Ginar at the time the bulk of the villagers moved, but returned to Gimurugi, while two other households (Nos. 5 and 11) built two dwelling houses each at Ginar and Gimurugi. The most plausible reason is that their sago and coconut groves were concentrated around Gimurugi.

When moving to a new village settlement, each household builds a small house, about two or three m by three or four m in size. Staying there for a couple of years, they build a proper dwelling house, which is rectangular, usually five by eight m (Photos 3 and 4), although, during my study period, a household with a *miid* man (Indiv. No. 6-1) and his two wives remained in two of the above-mentioned small-type houses for over four years. And a widow's (Indiv. No.

19-4) house, which was built by her sons living in Iamega village, was small, less than four m by five m. For both types of house, the floor is raised 1.5 m or more from the ground like a pile house and is supported by tough poles (made of trunks of *Planchonia papuana*, *Euodia* sp., etc.) at intervals of about 1.5 m lengthwise and breadthwise. These poles appear to last for over a decade. The space under the floor is not walled and is frequently used by the women in making baskets and mats and by the men for making bows and arrows, and for cooking, chatting and resting in the daytime. The floor is made of hard trunks of wild palms (*Ptychosperma* spp., etc.) that are split into two or three convex-shaped pieces. The roof is thatched with sago pinnae that have to be replaced frequently. Walls are made of rachises of sago palm, which are split into pieces around two cm in width and then woven crosswise; in a few cases, sago pinnae or tree bark (*Melaleuca ?leucadendron*) takes the place, to some extent, of sago rachises. The rachises usually last for a few years, while the endurance of sago pinnae is much shorter. There are usually some open spaces on the walls that serve as a kind of window, but the inside of the house is fairly dark even in the daytime. Some houses are equipped with a fireplace, which is not fixed on the floor but is an independent, portable structure made of steel plate or other steel materials, which have recently been introduced.

Besides the dwelling house, the villagers usually build huts (Photo 5) in their big gardens, even if these are located close to the village. Also they sometimes make huts near the big sago groves, which are usually more than one hour's walk away. Normally, the floor of such huts is not raised and the space between the floor and the roof is narrow. Sometimes there is no floor and it is open-sided. Both kinds of huts can be used for lodging, but normally the garden hut is used for resting in the daytime only. These huts are not usable after a couple of years. When the people stay away for a couple of nights for sago-making or other purposes at places without a hut nearby, they will make a very simple shelter consisting of a one-sided oblique wall covered with tree bark and/or palm leaves (Photo 6). During overnight stays when the weather is fine in the dry season, the people will sometimes sleep on the ground in the woodland or in the savanna, only making the spot even by clearing trees, shrubs and roots. All the sleeping sites, from the open sleeping ground to a village house, are called *met.*

Villagers' Producing Areas

Among food-getting activities, procurement of wild resources is freely pursued by all villagers without special advance care. Sago palms, coconut palms and garden crops go along with land ownership. Sago palms and coconut palms are owned by a married male, while garden crops are usually owned by a household or a husband and his wife or wives. Furthermore, due to their long maturation time, the former two plants are sometimes inherited through generations. Sago groves in swamps, which produce the bulk of the villagers' sago flour, might be used permanently. On the contrary, garden plots exist over a wide area of the savanna woodland and small forest along creeks or in savanna, depending upon the yearly selection. Seeing the village-land as the villagers' activity space, the distribution of these producing areas provides essential information about the environmental setting.

The spatial distribution of each household's sago groves, coconut groves and gardens under use between December 1971 and February 1972 is shown in Figure 11, in which sago and coconut groves are indicated by different marks, according to whether the grove was inherited from ancestors or was begun by the present owner. These locations were identified by the place names of the sites which were provided by the owners in accordance with my request, although most of the locations were actually visited by me.

An overall comparison of these figures characterizes the contrast in distribution between scattered sago and coconut groves and concentrated horticultural gardens. The variation in the number and location of sago and coconut groves among the married males is also prominent. The number of sago and coconut groves is largely associated with the number of groves owned by his father and the number of his brothers, both of which will determine how many groves and stands he will inherit. Two household heads, Indiv. Nos. 6-1 and 7-1, who transferred from the villages of Podare and Zim respectively after or at the time of marriage, possess no inherited groves in

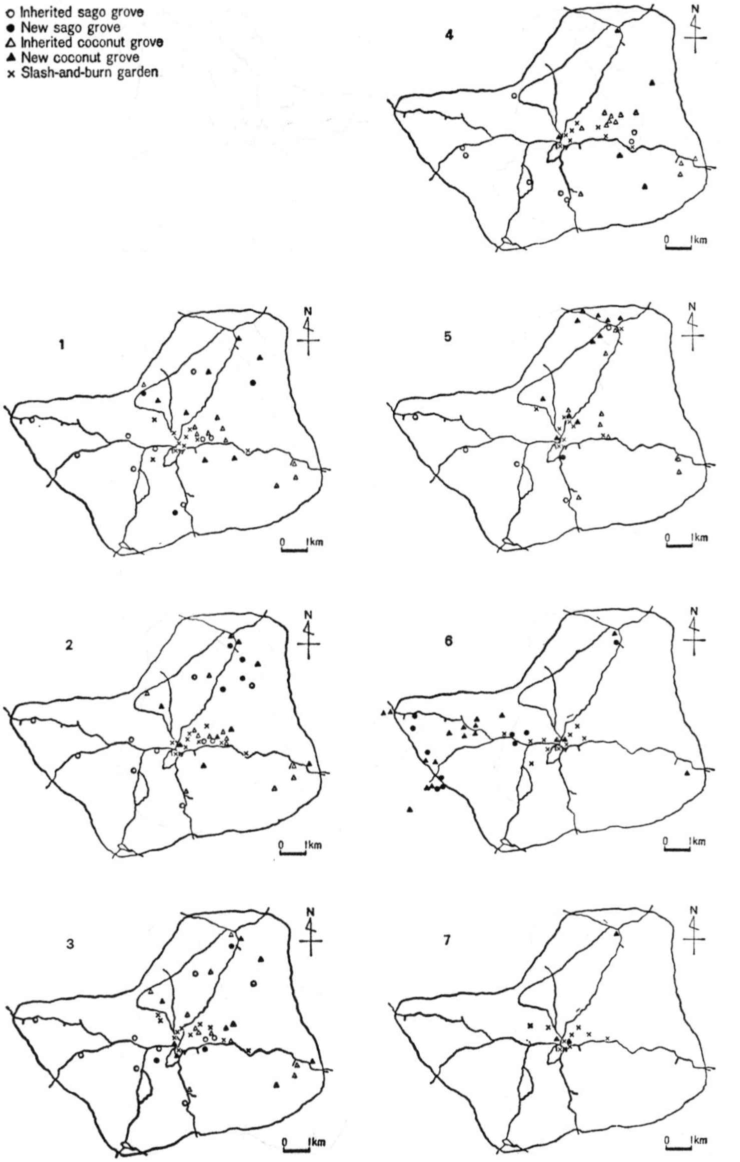

Fig. 11. Spatial Distribution of Each Household's Sago Groves in Swamp, Coconut Groves and Slash-and-Burn Gardens (under practical use in December 1971 to February 1972).

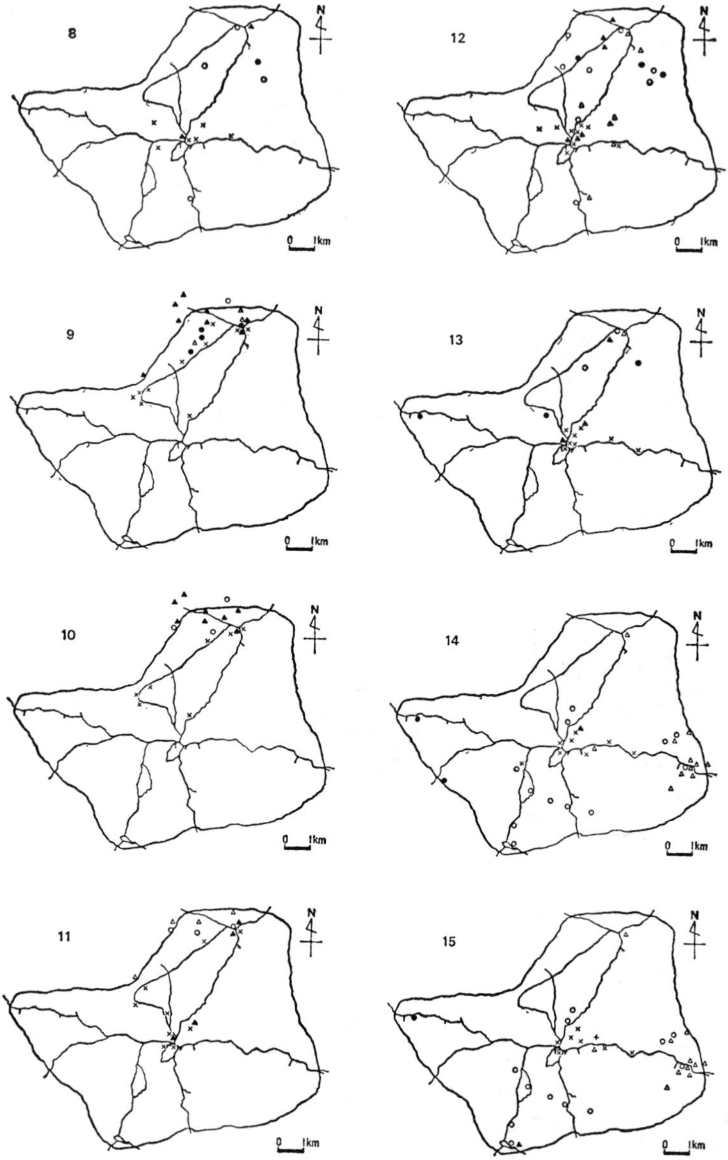
8
12
9
13
10
14
11
15
N
0 1km

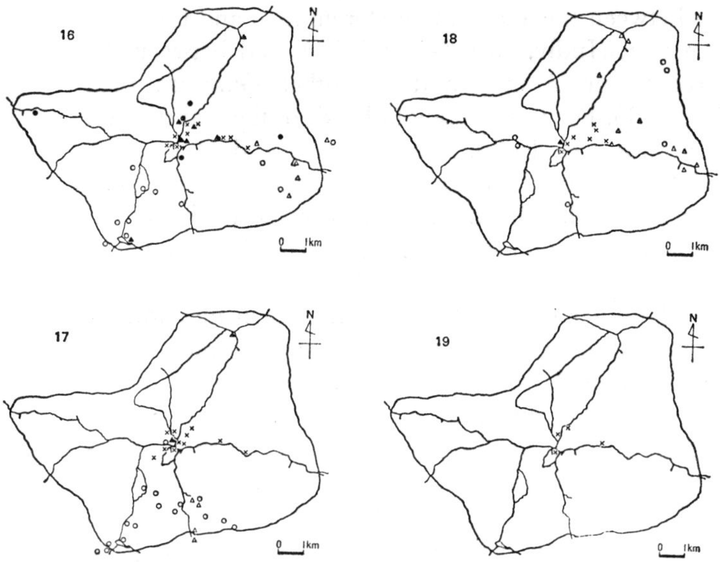

the Wonie village-land, although the household head from Gamaeve village, Indiv. 17-1 possesses inherited groves in the southern part, which is close to Gamaeve.

Aside from the in-migrants, the distribution of these groves reflects to a considerable degree the main activity area of each owner's ancestors. According to the villagers, in olden times (perhaps several decades ago) the Gidra people did not always settle down in a large settlement like the contemporary village, but congregated in several households belonging, as a rule, to a single family through the male line. Therefore each family kept a smaller area of central activity ground, even if such grounds overlapped one another. This situation greatly contributes to the diversified contemporary distribution, but at the same time contributes to a similar distribution pattern among owners who are brothers or belong to the same clan, especially to the same family group. The following explanation is coupled with the profiles of the households, as mentioned previously (pp. 47–50).

Good examples of similarity are seen in the distributions for the three household heads of the Dorem clan (Indiv. Nos. 1-1, 2-1 and 3-1), and those of the three heads of the Gameam clan (Indiv. Nos. 14-1, 15-1 and 16-1), although there are greater differences between the above three Dorem men and the other two Dorem men (Indiv. Nos. 4-1 and 5-1), and between the above three Gameam men and the other Gameam man (Indiv. No. 17-1). For five household heads of the Dantungam clan, two groups, i.e., Indiv. Nos. 9-1, 10-1 and 11-1 as one group and Indiv. Nos. 12-1 and 13-1 as another, are sibling and/or father-son related, and thus have respectively similar distributions of groves; nonetheless, the distribution as a whole is biased to the northernmost area, which is the Dantungam clan-land. The other four household heads, each individually belonging to a different clan, own sago and coconut groves differently distributed from each other. Most groves of the two men of the Wammopam clan (Indiv. No. 6-1) and Kameam clan (Indiv. No. 8-1) tend to be close to their respective clan-lands within the Wonie territory. On the other hand, the Rayaiena man (Indiv. No. 18-1), whose clan-land is located far away, possesses the groves in the eastern part, most of which were begun by his father who had migrated to Wonie. As already mentioned, a migrant of the Rekam clan (Indiv. No. 7-1) has planted a few new groves in the central part.

In this respect, the reason why two households (Nos. 9 and 10) stayed in Gimurugi may be partly explained by the concentration of their producing areas. Also noticeable is the fact that some household heads like Indiv. Nos. 6-1 and 18-1 left their clan-lands to venture out into an exploitation of their own main activity area.

There is another general tendency in the spatial distribution of each owner's sago and coconut groves. The location of new groves tends to be close to the village site and/or in the opposite direction of the area of concentrated inherited groves. However, such new groves usually contain a small number of matured stands because the construction of a new grove normally advances gradually; that is, the planting involves one or a few sago shoots or coconut seedlings. Furthermore, the expansion of sago groves is regulated to a considerable degree by the environmental setting. As mentioned previously, since the natural swamps have already been occupied by sago palms, the new groves are located in places along small, seasonally dried-up creeks which do not accommodate many stands. Thus the importance of the inherited grove, in which

Table 12. Matured coconuts possessed by each married male

Indiv. No.[a]	age-grade	inherited groves		initiated groves		total	
		no. of groves	no. of stands	no. of groves	no. of stands	no. of groves	no. of stands
6–1	*miid*	0	0	18	88	18	88
1–1		12	158	6	32	18	190
9–1	*nanyuruga*	2	4	10	106	12	110
10–1		0	0	8	60	8	60
5–1		9	112	11	73	20	185
12–1	*rugajog*	6	54	7	28	13	82
4–1		14	206	3	12	17	218
2–1		8	16	5	26	13	42
7–1		0	0	3	6	3	6
11–1		5	33	3	8	8	41
17–1		4	42	2	12	6	54
3–1		10	30	3	13	13	43
16–1		7	100	5	16	12	116
13–1		1	16	3	11	4	27
18–1		10	61	1	7	11	68
14–1		6	31	1	2	7	33
8–1		0	0	2	10	2	10
15–1		5	30	1	1	6	31

[a] Married males are arranged in decreasing order of age.

new stands have been added not only by natural propagations but also by transplanting shoots or seedlings, is very great.

The importance of these palms for subsistence should be assessed more by the number of stands than by the number of groves. The matured stands of coconut owned by each married male is shown in Table 12, the total number in the village-land exceeding 1,700. But, in spite of sago's greater importance for subsistence, I was unable to count sago stands because of their much greater number, the difficulty in identifying the palm's maturity (in contrast, coconut palms are easily identified by whether they bear fruit or not), and especially the greater inaccessibility of the inside of the swampy sago groves due to the limited survey duration. Judging from my limited observations, it seems that the variation in the number of coconut stands among individuals is roughly proportionate to that of the sago stands among them. Throughout the village-land the number of matured (plus semi-matured) sago palms is double that of the mature coconut palms. In other words, about 3,000 mature sago palms might exist in the village-land.

Another possible estimation of the number of sago stands comes from food consumption records, which state that three sago stands with full starch content are utilized per person annually. If unsuccessfully utilized trees were taken into account, this number would reach five. In this estimation, 500 trees would be cut or wasted in a year within the whole village. If sago palms in the last several years of their maturing process are regarded as "matured," the above estimated number of palms would be approximately accurate.

The number of slash-and-burn gardens and their distribution is almost identical among all households, except for the Gimurugi-dwellers, particularly the two households (Nos. 9 and 10) which have no dwelling in Ginar. The gardens are fairly concentrated around the village, and even the Gimurugi-dwellers tend to make some gardens close to the main settlement of Ginar. Though not so obvious, the distribution of garden plots for each household tends to be in the direction of the head's sago and coconut groves.

Based on the differences among both households and individuals in the distribution of these producing areas, two essential environmental situations have been observed. First, in spite of the nature of collecting in terms of actual activity pattern, the exploitation of sago and coconut are associated with a land-tenure system, which is popularly seen in most agrarian societies in New Guinea (e.g., Crocombe 1972) or in the rest of the world where land-tenure is commonly and intensively applied to horticultural plots. Second, contrary to horticultural systems of many other societies, slash-and-burn plots of the Oriomo are freely selected by each household and the space for gardens is not inherited. Thus most gardens are concentrated around the village and the inter-household differences in distribution are of lesser importance.

3. Life in and out of the Village

Village Life

The Wonie day begins around sunrise, at approximately 6:00 (time given in this volume follows Papua New Guniea

Standard Time, which is adjusted for the longitude of 150°E). The villagers come out of their houses when visibility is barely from one end of the village settlement to the other, or about 200 m. An exception is sometimes seen in cases where adult males set out individually to hunt before daybreak. The fairly synchronized awakening time (or, precisely, the time at which distinct activity commences) is due to a societal contempt for oversleeping. However, the people will stay in the village for a while after rising, and the onset of distinct activity varies greatly according not only to age and sex but also intrapersonally from day to day. Walking in the village, sitting under or in front of the house, chatting with others, making a small fire for warmth (by aged persons) and cooking (usually by women) are involved in a typical morning scene.

The synchronized awakening time does not necessarily correspond to any synchronization of time in going to sleep. The people usually retreat into their houses by 20:00 or 21:00, but there are many matters which can interfere with this. Dancing is held in the village several days a month and most villagers and visitors, if any, attend. The dancing party usually begins in the late evening and continues until midnight or, sometimes, even to daybreak. Villagers participate in dancing to the extent that they wish. The villagers join the dancing circle individually or in twos or threes, and retire from it at any time they like. Some other events of great interest, such as a quarrel and/or fighting, a person's illness or injury, a childbirth, or a person's death, can make the people keep late hours. Besides the events mentioned above, chatting about various topics sometimes continues accidentally far into the night. A person, especially a male, who slept for little or no time during the night may take a nap the following day.

Activities performed within the village settlement do not involve direct food production because the villagers do not raise animals, and their gardens and other producing areas are located outside the village. Another characteristic is the lack of routine daily work except for cooking meals and fetching water from the waterplace, several minutes' walk away. The cooking is done two or three times a day; the details will be explained later. The villagers jointly use a permanent waterplace, which is the uppermost part of a creek. It is divided into two sections, the upper for drinking water and the lower for bathing. This small creek does not contain water in any dry season. In 1971, when my main fieldwork was carried out, the waterplace did not go dry. In 1980, however, the water did dry up, and water was then sought from various small swamps or segregated creeks within half an hour's walk from the village. The people like bathing, and it is done once or twice per day in fine weather. When returning from hunting, gardening or sago-making, the people tend to have a bath in creeks close to the village, while their regular bathing is usually done at the village waterplace.

Much time in the village is spent at leisure. Chatting, resting and caring for or playing with children occupy the bulk of it. Nonetheless, the making of bows and arrows by males and several types of baskets and mats by females demands considerable time in village life. On the average, several days per month are used for such activities by each individual, while daily work time is two or three hours. Usually the people do not seriously devote themselves to such work but rather do it at an arbitrary pace while chatting or so on. Another village-centered activity is to collect firewood; this is done fairly often but not every day. Firewood is the only fuel for this society and contributes to cooking, lighting the inside of the dwelling house, warming the body while sleeping, illuminating the dancing circle in the village yard, etc. The people's knowledge of a variety of woods, barks and branches is so rich that the firewood is divided into at least four categories, initial-firing, briskly-burning, persistently-burning, and lighting. Most firewood is gathered, usually by housewives, from both old and new horticultural gardens or from the surroundings of the village.

Quarrels sometimes occur in the village, and some of them develop into fighting. Most frequently occurring are quarrels between a married couple, although the frequency varies greatly from couple to couple. The origin of these quarrels is usually very trifling, but the scale of the quarrel may not depend on its origin because more and more new themes tend to be invented. The final theme in cases which develop into fights usually concerns the wife's non-pregnancy or lack of offspring.

Quarrels involving more than two adult males, married and

1. The Oriomo River

2. The village of Wonie

3. A large house in the village

4. A small house in the village

5. A hut in the garden area

6. Shelters

kewalbuga, occasionally bring about large-scale fighting, although the direct causes of these seem to be insignificant as well. Popular themes for such quarrels derive from the recollection of events in the past, regardless of the initial themes. As almost all males in the village come to take part, the quarrel results in fighting between two groups of males, corresponding to their moiety. One group involves males of the Dorem clan, and the other a dual alliance of Dantungam and Gameam clans, both belonging to the Dantungam moiety. Other males usually stand by the group of their moiety and sometimes actively take part. Women, in groups, surround the fighting spot at a distance, while some denounce the fighters of the opposing moiety. To excite themselves for village-scale fighting, they sometimes drink juice squeezed from the stems of two species of *Piper* grasses (a natural growth of *Piper umbellatum* called *ukuk* locally, and another *Piper* sp. called *uk*, which was recently introduced). My experience in testing the effects of *uk* juice has proven very similar to the kava (*Piper methysticum*) drinking reported by Holmes (1979): "a refreshing, astringent drink which produces nothing more than a tingling sensation in the mucous membrane of the mouth and a short-lived numbness of the tongue," although the *uk* drink was associated with a "bitter" rather than a "refreshing" taste. According to my observations, this type of quarrel associated with fighting took place once every two or three months. The most severe case involved almost all the adult males of the village and lasted until midnight when two of them were knocked down, having been smeared with blood.

Matrimonial quarrels, if they involve fighting or throwing objects, do not usually result in long-term hostile feelings between the married couple; even by the morning of the following day the relations of the couple will be restored. On the other hand, the village-level quarrels occasionally produce a conflict among the people, especially among those who hit each other, but it lasts only several days at most, even if the ill-feeling remains in concealment. It can thus be concluded that the quarrels among the Wonie people do not produce enduring hostilities to the extent to which their daily lives are profoundly affected.

The recent introduction of Christianity has acquainted the people with the meaning of Sunday, but their comprehension of the conceptual difference between the days in a week is still vague. Precisely speaking, the villagers know the meaning of Sunday, but they have prayer meetings on Sundays very sporadically and usually carry on their food-getting activities. The fact that no church existed in the village and no pastor lived there did not make them very Christianity-minded. Pupils of the primary school at Wipim, less than ten in number, did not always go to school, but when they did they normally came back to the village every other Friday and went back on the following Sunday or Monday. However, whether the boys and girls stayed in the village or at the school did not greatly affect the adult villagers' activities. Hence, it can be said that a weekly cycle has begun influencing their life, but the people's activity pattern remains almost unchanged.

Overnight outside the Village

The village-land is, as a rule, a whole entity supporting the Wonie people's subsistence. The villagers do not always spend their nights in the village settlement, but sometimes stay overnight at other places within the village-land for food-getting activities. Also, they occasionally make calls at other villages and at the town of Daru and stay at those places. These overnight activities are regarded as a part of their daily activity pattern, and thus the destinations and frequencies of such visits are necessary for an understanding of the framework of their activity system. Furthermore, the visits contribute to an identification of individual differences in activity pattern on the one hand, and to an understanding of relationships with the adjacent villages on the other.

The relevant data were gathered over three periods totalling ninety-two days: one, from 28 July to 17 August 1971, during the semi-dry season; another, 14 October to 5 November 1971, during the dry season; and the last, 10 February to 14 March 1972, during the wet season. Those who were observed in the surveys were exclusively adults belonging to the *miid*, *nanyuruga*, *rugajog* and *kewalbuga* stages for males and the *nanyukonga* and *kongajog* stages for females. Table 13 shows the frequencies of visits to the eight categories of overnight places. In this analysis, the villagers are classified into two groups,

Table 13. Percent distribution of overnight places

A) Ginar-dwellers

age-grade	total person-days of observation[a]	Ginar	sago groves	gardens	hunting	community work	Gimurugi	other villages	Daru
miid	184	84.2	8.7	3.3	0	0.5	0	3.3	0
rugajog	993	66.8	9.2	1.3	0.2	4.3	0.3	9.8	8.2
kewalbuga	429	43.4	0.7	0	0.2	10.5	0.5	5.6	39.2
nanyukonga	368	75.5	10.9	4.9	0	0.5	0.3	7.9	0
kongajog	1,254	75.4	9.6	2.1	0	3.2	0.2	7.2	2.3

[a] Due to in- and out-migrations, some villagers were not fully observed.

B) Gimurugi-dwellers

Household No.	status and age-grade	Gimurugi	Ginar	community work	other villages
9	head: *nanyuruga*	90.2	9.8	0	0
	wife: *nanyukonga*	90.2	9.8	0	0
10	head: *nanyuruga*	69.6	28.3	2.2	0
	wife: *kongajog*	70.7	29.3	0	0
11	head: *rugajog*	56.5	34.8	5.4	3.3
	wife: *kongajog*	58.7	41.6	0	0
5	head: *nanyuruga*	38.0	52.2	2.2	8.7
	wife: *kongajog*	35.9	55.4	0	8.7

one with a dwelling house only at the Ginar main settlement and the other with a house at Gimurugi, the latter involving four households, two of which possess houses at both settlements. The latter group's activities differed from those of the former group, and in fact they could not be observed in many cases simply because I stayed in Ginar.

The categories of "sago groves," "gardens" and "hunting" in Table 13 imply cases when the villagers stayed overnight for these food-getting activities within the village-land. The community work which took place during my study period included the construction of a wide road extending east and west along the central zone of the Oriomo Plateau, and the repair of the primary school house and its associated dormitories in the vicinity of Wipim village. The voluntary, unpaid school repair was made in cooperation with people from several villages. The road construction project was promoted by the government of the Western District (on behalf of the Central Government), although the actual labor was organized by the villagers through the efforts of a councillor of the Oriomo-Bituri Local Government, who is a *rugajog* man (Indiv. No. 2-1). The workers received cash payments, though very small, once a year. The major work in road construction involved cutting trees on the site with axes and making side-drains with shovels. The lodgings for the road construction project were usually within the village-land.

Visits to other villages usually take place with the aim of meeting relatives and friends, and this typically occurs between persons who are parent-child-related, siblings or partners in an exchange marriage. Occasionally gifts are exchanged, the most typical being food crops. Serious illnesses or deaths also spur inter-village calls. Mutual visits among the neighboring villagers contribute to the information flow as well, but seem to have little direct economic involvement. One exception among the Wonie villagers was the visits to Zim village made by a couple of Household No. 7 whose head had transferred from Zim several years ago and who still utilized his sago palms in the Zim land. Judging from the infrequent overnight visits to Gimurugi by the Ginar-dwellers, it seems that the Gimurugi settlement was regarded as similar to other adjacent villages. However, the Ginar people did make occasional one-day visits to Gimurugi, while one-day visits to other villages were very seldom made. Visits to Daru, two or three days' travel each way, involved the councillor's participation in the assemblies held there several times a year and visits to the Maranatha Mission by various villagers, especially *kewalbuga* males, at its request. In addition, there was one lengthy visit by Indiv. No. 4–1, who was employed in paid work in Daru. He spent about three months there, although the work was available to him only intermittently, a couple of days a week on the average. Any visits to Daru were usually coupled with the purchase of various necessities for all the villagers. On the other hand, the things to be sold in Daru included mats made of pandan leaves, small-sized carvings of human faces and/or animals like cassowaries and pigs, although these were not very large in number nor were they always brought along.

The frequency of overnight visits reflected the villagers' life style, and a considerable difference existed according to sex and age-grade, for the following reasons.

Sago groves and gardens. More than ten percent of the married members were involved in overnight activities for sago-making and gardening despite the fact that most working sites were located within one hour's walk from the village. This may suggest semi-permanent utilization of the village settlement in this society, thus differing from "proper" horticultural societies. The inter-group differences in such activities were specifically characterized by the low participation of *kewalbuga* males in contrast to the high proportion of married villagers. While more detailed reasons for this will be given in later chapters, the most basic cause is the *kewalbuga*'s minimal involvement in all actual subsistence activities except hunting.

Community work. The high participation of *kewalbuga* males is connected to the scant time they spend on subsistence activities, and in fact they tend to participate in all community work as long as they are in the village-land. This is because they are very eager to get cash from road construction work, and also because this age-grade has always been traditionally regarded as responsible for contributing to incidental activities on the community level.

Other villages. The inter-group variation in this category is relatively small. Despite the fact that *miid* men showed the lowest proportion of participation, they customarily make visits for serious events, like human death. The higher proportion among the remainder seems to be attributable to their greater physical strength, which enables them to make frequent visits in order to enjoy themselves, as well as to make non-specific contact with the people of other villages.

Daru. The visits to Daru vary to the greatest extent from group to group. The pattern is similar to that for community work. Because they are not involved in everyday food exploitation and because of their adventurous spirit, *kewalbuga* males make more visits due to their increased cash income and the greater availability of lodging at the mission's dormitory. The large portion of visits by *rugajog* reflects the councillor's participation in the assemblies and Indiv. No. 4-1's long stay, as mentioned previously.

The proportional allocation of overnight activities for the four households of Gimurugi-dwellers for community work and visits to other villages as shown in Table 13 involved only those times when the people left or returned to Ginar. In other words, the records did not distinguish Gimurugi-centered overnight activities to places other than Ginar. Aside from this, the rate of traffic between Gimurugi and Ginar reflects the degree of each villager's involvement in the two sites. The members of the two households (Nos. 5 and 11) which had dwelling houses at both Ginar and Gimurugi settlements stayed at each place almost equally.

A breakdown of the Ginar-dwellers' overnight visits to other villages is shown in Table 14. The villages visited during a ninety-two-day period for this study included five out of

Table 14. Villages visited and stayed in overnight by the Wonie villagers

village	total no. of Wonie people who visited	total no. of person-days (%)
Zim[a]	2	86 (35.0)
Wipim	22	78 (31.7)
Gamaeve	9	33 (13.4)
Iamega	4	19 (7.7)
Ume	3	17 (6.9)
Kuru	6	13 (5.3)
total	46	246

[a] The visits of the two villagers (a married couple of Household No. 7) included sago-working, although I did not determine the proportion.

the six neighboring villages, i.e., Zim, Wipim, Gamaeve, Iamega and Kuru, and one village, Ume, located about twenty km away. This list of villages suggests the villagers' narrow sphere of communication. The village Zim was most frequently visited but by only two villagers, i.e., the couple of Household No. 7, because, as mentioned previously, their visits involved several sago-making trips. Aside from the village Zim, the total number of visitors roughly correlates with the total number of days of overnight stays. The inter-village difference in frequency almost corresponds to the number of people with whom the Wonie villagers are related by kin or affinity.

3. Subsistence Activities

1. Food-getting Activities

The Oriomo Papuans maintain their subsistence by various food-getting activities which include hunting, fishing in small creeks, collecting wild plants and small animals, utilization of sago starch, utilization of coconuts, and slash-and-burn horticulture. One of the most striking features in the above combination is the lack of animal husbandry, but hunting frequently provides animal food. If plant-food-getting activities are, in general terms, grouped into the collecting of wild resources and the cultivation of domesticated resources, the utilization of sago and coconut among the Oriomo Papuans combines both these aspects. Transplanting shoots of the sago palm is frequently practiced, as is that of coconut palm seedlings. Although these palm stands are sometimes visited by the owner, no special preparatory labor is required before the actual harvest, as is true in collecting wild food. Thus, plant food procurements can be divided into three groups: collection of wild plants, exploitation of sago starch and coconuts, and garden horticulture.

While each food-getting strategy will be delineated, it should be remembered that all coexist as complementary parts of a single subsistence pattern, and that no individuals or households are specialized into any of the activities as a social function. What I intend to delineate in this chapter are the descriptive aspects of each food-getting activity, including the associated implements, and reports of some typical cases which I observed thoroughly.

Hunting

Hunting is practiced exclusively by adult men; I observed women capturing rats only once, as described in Appendix 5. While *yambuga* boys play with bows and arrows and sometimes aim at small animals like lizards and small birds in the village and the slash-and-burn gardens (Photo 7), or infrequently accompany the adult males, especially *kewalbuga* males, on hunting trips, their contribution to food procurement is completely negligible. It is at the *kewalbuga* stage that a man starts to become an active hunter. As a man grows older, that is, as he enters the *nanyuruga* stage and then the *miid* stage, he tends to retire from hunting. One (Indiv. No. 1-1) of the two *miid* men in Wonie gave up his bow and arrows and retired from hunting completely. While he was said to be two or three years younger than the other *miid* man (Indiv. No. 6-1) who still had his bow and arrows, the former was much inferior physiologically, judging from my measurements of grip strength and visual acuity (see Fig. 7). The latter *miid* man and three *nanyuruga* men, all of whom were physiologically inferior to the *rugajog* and *kewalbuga*, had their own bow and arrows and took part in many communal hunting expeditions, but they actually caught few game animals throughout my study period.

The bow for adult men, which is about twenty cm longer than the body height of its owner, is made of a kind of bamboo (*Bambusa vulgaris* Schrad.), and the width and thickness of the stave at the center are 4.5–5.0 cm and 2.0–2.5 cm, respectively. The bowstring, which is about five mm wide and less than one mm thick, is also made of *Bambusa vulgaris* (Photo 8). When a male becomes a *kewalbuga*, he will usually learn how to make a bow from his father, his *omugam* or some other adult man. The bow maker carefully selects a bamboo cane to bring to the village. After it dries, he begins scraping it with a steel knife (or freshwater shells in former days). This process progresses little by little for several weeks, during which the bamboo is frequently soaked in water and dried on the fire. Finally, the surface is abraded with the sandpaper-like leaves of *Ficus ?opposita* and polished with several kinds of tree sap. Many active hunters made their own bows, although some who were said to be unskilled did not take part in bow making. Such unskilled persons were presented bows by other villagers. Actually, the gift of a bow is fairly common in this society, especially among kin- or affinity-related males, without necessarily regard to bow-making skill. The people say that on the average two or three bows of an active hunter are spoilt in a year.

Arrows used in current hunting practice are divided into four types, depending on the shape and material of the arrowhead, while all arrowshafts are made of bamboo grasses, usually *Bambusa atra* Lindl. For assembling the arrowhead and arrowshaft, the shaft end into which the arrowhead has been screwed is wound and bound with threads made of the inner bark of the tree *Gnetum gnemon*. This binding is done while the adhesive sap of trees (usually *Glochidion* sp. or *Schizomeria* sp.) is smeared continuously over the threads. Both *gnalib*, with a slender-pointed arrowhead made of the trunk of the wild palm (*Ptychosperma ?macarthuri* or *Hydriastele* sp.), and *girisung*, with an arrowhead made of an attrited steel bush-knife, are used for all terrestrial animals including cassowaries of the Ratitae. Actually, *girisung*, a new type of arrow, has normally been used for the past few decades because of its greater power. *Dupa*, with a blunted arrowhead made of hard wood such as *Halfordia papuana* Laut., is used for hunting small birds, although no birds except the cassowary are on the principal game list of the Oriomo Papuans. The last type of arrow, called *waia* and introduced perhaps as recently as *girisung*, has a multi-prong of iron wires and is usually used for fish and, rarely, for small birds and fruit-bats. Historically, *girisung* took the place of *sung* whose head is made of a sharpened stalk of *Bambusa vulgaris*, although many villagers still possess some *sung* arrows even if they are not often used. There is a great variety of *gnalib* and *sung*, depending on the shape and decoration on the arrowhead.

The bows and arrows used by the Wonie villagers at present and in recent years are referred to in detail in Watanabe's (1975) report. From his study and from my own observations and measurements, the quantitative analysis of their arrows is briefly as follows. The length of the arrowheads used by the adult hunters averages about 40 cm for *gnalib*, 20 cm for *girisung*, and 10 cm for both *dupa* and *waia*, while the total arrow length ranges from 90 to 170 cm, with *gnalib* and *waia* relatively longer and *girisung* and *dupa* shorter. The weight of these arrows varies greatly from 60 to 200 g but most are between 80 and 140 g, according not only to the type but also to the owner. However, individual differences in the weight of arrows do not seem to correspond directly to the owner's physical strength but rather to depend on casual reasons like the nature of available materials. When arrows are spoilt because of damage to the arrowhead, a break in the joint between the arrowhead and arrowshaft, etc., the shaft as well as the head are made re-usable through shortening as far as possible. The shafts and heads other than *girisung* are given up when they become three-fourths to four-fifths of the original length, while *girisung* heads seem to be used for several years or more. It is assumed that each active hunter makes ten to twenty arrows every year, although gifts of arrows sometimes occur.

Several years ago, a shotgun was first introduced to the Wonie village, and three have been owned by three different persons (Indiv. Nos. 2-1, 11-1 and 18-1). The superiority of the shotgun to the bow and arrow is evident in individual hunting for such birds as pigeons perched in tall trees and also for shooting big animals like cassowaries and crocodiles. On the other hand, the shotgun is not desirable for group hunting in which the hunter aims at an animal which runs out unexpectedly from within a fiery circle, since the preparation time for shooting with a shotgun is longer than that for bow and arrow. Individual hunting accompanied by dogs is impossible using a shotgun. Thus, the merits of the shotgun are offset to a considerable degree by these demerits. Furthermore, shotgun bullets, manufactured in urban areas, cannot easily be purchased because of the money and transportation involved. The gun owners own bows and arrows and use them frequently as well. Consequently, animals killed with shotguns occupy only a small percentage of the total catch in Wonie.

Hunting strategy is principally divided into two types: individual hunting and group or communal hunting, the former using bows and arrows or shotguns as weapons, and the latter bows and arrows exclusively. Besides these, deadfall trapping, with strained draff of sago pith as bait for feral pigs, is occasionally practiced. During my study period, only two deadfalls were made and only one pig was killed. The villagers say that their traditional strategies include another type of deadfall for cassowaries, in which the cassowary's favorite fruits are used as bait, although I did not witness that construction. Another strategy, for capturing pythons, is to have the men thrust their sticks, manufactured on the spot, into the animal while it lurks in the mounted nest of a kind of bush fowl (*Megapodius* sp.). The nest is about one m in height and several square m in area. This method is possible for a short period during the early dry season when the bush fowl's eggs lie at the bottom of the nest. During my observation, however, this strategy succeeded only twice.

Although the vegetation of the hunting ground differs according to the game animal hunted, the distance from the village may not play a decisive role in the selection of hunting grounds. The spatial distribution of hunting grounds covers a wide area of the village-land.

Individual hunting is practiced throughout the year both in the savanna and in the forest. The hunter goes hunting by himself, sometimes accompanied by hunting dogs. About a dozen adult dogs were kept by the villagers during my study period. Although the dogs have not been specifically trained by the hunters, they may have learned how to chase game animals while imitating other dogs' behavior in hunting sessions. The ownership of each dog is clearly recognized, but when hunting with dogs it is normal for a hunter to take several dogs, most or all of which are not his own.

A usual arrow set consists of from one to three *girisung* and several *gnalib*, and often a *dupa* and/or a *waia*. The main game animals (Table 15) are cassowary, bush-wallaby and scrub-wallaby in the forest, and grass-wallaby and bandicoots in the savanna. Feral pigs, which are also among the most important game, inhabit both vegetation zones. Table 15 also lists the game animals of secondary importance. Lizards and birds smaller in size than pigeons are recognized as only a small part of the meat supply in spite of their abundance and the ease of obtaining them. However, larger animals, preferred for their size, are more difficult to find. The hunter's behavior in individual hunting varies to a considerable degree depending on the kind of game being hunted (Photos 9 and

Table 15. Game animals

A) Principal game

common name	local name	scientific name	average weight (kg)	vegetation inhabited
pig	boma, worum	*Sus scrofa*	35	all
(grass-)wallaby	seba	*Wallabia agilis*	15	savanna
(bush-)wallaby	giur	*Dorcopsis veterum*	5	forest
(scrub-)wallaby	suga	*Thylogale* sp.	5	forest
bandicoots	ruei	*Echymipera ?rufescens*	2	savanna
	girag	*Echymipera* sp.	2	savanna
cassowary	gigi	*Casuarius casuarius*	40	forest

B) Other game

common name	no. of species	scientific name
possums and cucuses	9	*Pseudocheirus* spp., etc. and *Phalanger* spp., etc.
birds	ca. 100	
monitors	8	*Varanus* spp., etc.
lizards	6	*Egerina* spp., etc.
snakes	14	*Liasis* spp., etc.
crocodiles	2	*Crocodilus* spp.

10). Generally speaking, cassowaries and pigs are the most difficult to kill (Ohtsuka 1975, 1978). The hunting methods used for the three most important game animals, i.e., cassowary, pig and grass-wallaby, are as follows.

It is very difficult for a hunter to locate a cassowary because of its fine visual and auditory senses. First, the hunter must search in a wide area where the cassowary's favorite fruits lie on the ground. When he thinks that a certain spot has recently been visited and will be visited again by the birds, he constructs a screen of wild palm leaves about ten m away as an ambush. Alternatively, in the mid-dry season he makes a similar screen near the waterpool where cassowaries are expected to bathe, and ambushes them in the daytime. The hunting dog is ill-fitted for cassowary hunting simply because of its bark. The shotgun is suitable for this type of hunting, although bows and arrows are also widely used.

Locating a pig, a nocturnal animal, may also be impossible because of its excellent sense of smell. Pig hunting usually requires the aid of hunting dogs. When the dogs begin to chase the game, the hunter follows their noises. The longer it takes the hunter to arrive at the scene of fighting between the pig and the dogs, the greater the probability of the dogs' defeat and/or the pig's escape. In a successful case, the hunter shoots arrows into the pig as it struggles with the dogs.

Grass-wallabies are killed by Wonie hunters in greater number than any other animal. The density of this animal in the area is much higher than that of the other principal game animals, although no systematic information on the population density of the animals was obtained. In comparison to the above two animals, the grass-wallaby is easier to hunt since the hunter can find it while carefully walking around the savanna in the early morning or evening, the feeding hours of this marsupial. Normally, in successful cases a grass-wallaby is not observed directly but its existence is confirmed by the unusual movement of grass around it. Next, the hunter must approach by walking slowly and then by creeping as close to the animal as possible, ideally ten m away. While approaching, the hunter may make a sound by beating his chest if he is in an erect or slouching posture, or alternately by beating the ground as he creeps, thereby attempting to imitate the habitual sound of the wallabies; it is said that they hit the ground with their tails. The approach is the most difficult part of wallaby hunting. As a matter of course, dogs are not used for this type of hunting.

Case record of an incident of individual hunting. This hunting trip was observed thoroughly by me. Because it took place in the latest stage of my study, my presence did not likely disturb the hunter's behavior or the animals unduly. The hunter aimed for grass-wallabies, although no animal was killed.

1) Date: 17 March 1972.
2) Hunter: *Kewalbuga* male, Indiv. No. K-2. He carried with him a bow and six arrows, two *girisung*, three *gnalib* and a *dupa*, and was not accompanied by dogs.
3) Route of hunting travel in approximate scale is shown in Figure 12.
4) Time:
   6:45 departed the village.
   7:25 parted from the main path (see the route map of Fig. 12).

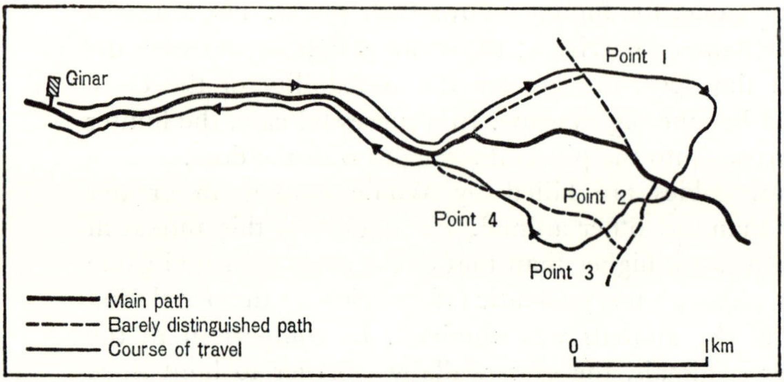

Fig. 12. Route Map of Hunting Travel (in approximate scale).

9:50 returned to the main path.
10:20 arrived at the village.

5) The hunter's behavior at four points (see the route map), where he perceived the existence of animals or witnessed grass-wallabies:

Point 1. When he walked into the savanna with thinly growing grasses (*Imperata cylindrica*), about 50 cm tall in many parts, a grass-wallaby suddenly appeared from 40 m away, where several trees and taller grasses (about one m in height) grew together. He did not do anything but watched it run away.

Point 2. In the savanna, with grasses 2 to 2.5 m tall, the hunter perceived the existence of an animal 30 m away from him by listening to the rustle of *Imperata* grasses caused by the animal's movements. At this moment, he might have recognized it as a grass-wallaby. Slouching, he carefully and slowly approached the animal while beating his chest with his right hand and holding a bow and arrows with his left hand. After a couple of minutes' advance of about 20 m from the starting point, the wallaby suddenly ran away while he shot *girisung* arrows without success.

Point 3. The environmental conditions at this point were very similar to those at Point 1. When the grass-wallaby unexpectedly moved away, there were only five to six m between the hunter and the animal, which might have been feeding in a small patched assembly of taller grasses. The hunter followed it about half a minute, but did not shoot any arrows.

Point 4. We noticed a sound from the right, and then saw a grass-wallaby run across into the savanna some 20 m beyond, where grasses of about one m in height grew together. The hunter pursued the wallaby but failed to follow it more than a couple of minutes.

Group hunting or communal hunting is practiced only in the dry season when the savanna grasses wither (Photos 11 and 12). For a group hunt, a party of from several to as many as twenty adult men is temporarily organized, infrequently accompanied by *yambuga* boys. The Wonie villagers have no permanent leader for group hunting. The group hunt is usually planned in the early morning of the day in question or on the previous evening after a proposal by one or a few of the hunters who, in many cases, had witnessed either animals or their footprints, and all of the active hunters currently in the village normally participate in the party. The members of a party arrive at the appointed spot either individually or in twos or threes. Then they discuss in detail the site of the hunt and their individual roles. Older persons or those who proposed the hunt tend to lead the discussion. One to four members, usually younger ones, set individual fires at a number of points by burning fronds of wild palms, thus making a circle about one to two km in diameter. The rest of the members stand around the circle at intervals, and those setting the fires later join them. Every hunter standing along the circle watches the grass inside the circle closely, holding a bow in one hand and an arrow in the other. Any abnormal movement of grass is a sign of an animal running away from the fire. Some of the running animals are caught by the hunters nearby, although a large number of animals inside the burning circle seem to be able to escape by passing through the spaces between the hunters.

The principal targets of group hunting are grass-wallabies and two kinds of bandicoot, while pigs and monitors are sometimes caught. Besides these, bush-wallabies and scrub-wallabies are killed when the burning circle contains savanna woodland. Generally speaking, group hunting is less tiring than individual hunting, and therefore the aged men tend to participate in the former but not much in the latter. *Kewalbuga* males also prefer to a great extent the former to the latter. Comparing group and individual hunting according to productivity (weight of game animals per hunting period) for *rugajog* men, who are the most skillful, the latter is slightly more productive. It should also be noted that, since any group hunting site can be used only once during a dry season, the hunters of Wonie cannot practice group hunting very often. During my study period, which covered the whole duration of the 1971 dry season, only seven group hunts took place.

Case record of an incident of group hunting. I observed and participated in this group hunt for its whole duration. However, it was natural that in practice my observation covered only several hunters, one or at most two at any moment. The group hunt in question resulted in a fairly large number of game animals.

1) Date: 15 October 1971.
2) Participants (Indiv. No. and age-grade in parentheses): 2-1 (*rugajog*), 3-1 (*rugajog*), 6-1 (*miid*), 7-1 (*rugajog*), 12-1 (*rugajog*), 13-1 (*rugajog*), 14-1 (*rugajog*), 15-1 (*rugajog*), 16-1 (*rugajog*), 18-1 (*rugajog*), K-3 (*kewalbuga*), K-7 (*kewalbuga*), K-9 (*kewalbuga*)
3) Location and vegetation: One and a half km north-northeast of the village, called Iejam locally. The major vegetation of the hunting ground was savanna, close to fairly large savanna woodlands as shown in the map (Fig. 13), while smaller patched woodlands grew within the fired circle.
4) Spatial arrangement of hunters: The hunters' approximate initial positions for the ambush are shown by their individual numbers in the map of Figure 13, although, naturally, they moved around frequently.
5) Time: Approximately from 12:30 to 14:00. The end of the group hunt was not certain, because some hunters gathered at the promised point while, simultaneously, some others continued the hunt.
6) Kinds and number of animals killed by each participant (Indiv. No., and age-grade in parentheses):
   2-1 (*rugajog*): pig (1)

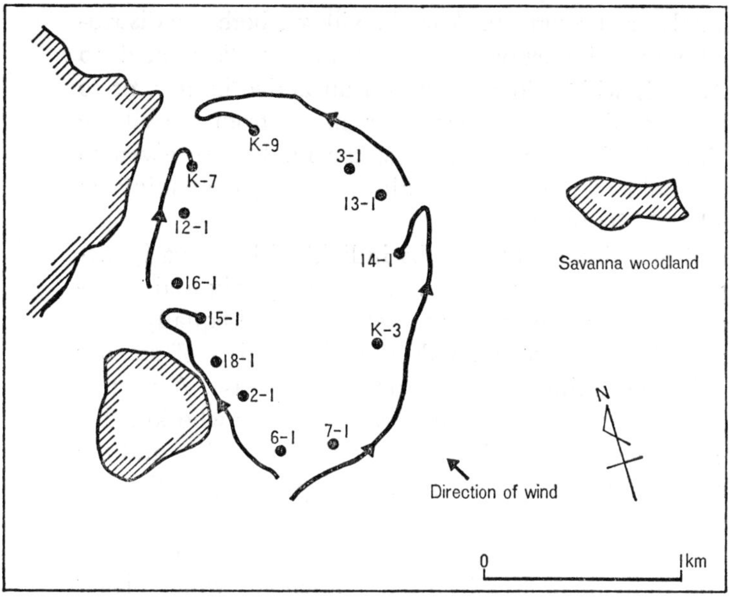

Fig. 13. Sketch Map of Group Hunting Ground (in approximate scale). The hunters' initial positions are indicated by their individual numbers. Four hunters (Indiv. Nos. 14–1, 15–1, K-7 and K-9) burned grasses, and their routes are also indicated.

3-1 (*rugajog*): grass-wallaby (1)
6-1 (*miid*): —
7-1 (*rugajog*): bandicoot (1)
12-1 (*rugajog*): —
13-1 (*rugajog*): bandicoot (1)
14-1 (*rugajog*): bush-wallaby (1)
15-1 (*rugajog*): bush-wallaby (1), bandicoot (1)
16-1 (*rugajog*): pig (1), bush-wallaby (1)
18-1 (*rugajog*): —
K-3 (*kewalbuga*): bush-wallaby (1)
K-7 (*kewalbuga*): bandicoot (1), small bird (1)
K-9 (*kewalbuga*): —

All animals killed by group hunting are, as a rule, first distributed at the spot after dissection of the bodies into large pieces, the equal share going to each married participant being much larger than the equal share going to each *kewalbuga*. Sometimes some portion of the catch is reserved for households whose members have not participated in the hunt.

The animals are brought back to the village, both the distributed and reserved portions, and are then redistributed to the households which do not contain an active hunter, or to active hunters who were either ill or absent from the village at the time. Whichever is the case, the final share given to each household of the village is almost equal, depending on the number of its members.

Disposal of animals obtained by individual hunts varies according to size. Bigger species, i.e., cassowary, pig and large-sized grass-wallaby, are usually presented almost equally to all households. On the other hand, the distribution of smaller species of game animals varies from case to case. A part of the animal is normally presented by the hunter who killed it to some households related to him by kin or affinity, and the number of recipient households varies, depending on the size of the animal.

Smaller animals, roughly less than 15 kg in weight, are cooked on individual fires by each household, while bigger animals are cooked by many people in a cooperative fashion, by baking the dissected animal meat with bones on burnt small pieces of pisolitic laterites or broken pieces of termitaria covered with banana leaves and the bark of *ti* trees (*Melaleuca* spp.); the details will be described later.

Fishing

In comparison with hunting, fishing brings much less animal food to the Wonie villagers. There are three types of fishing: (1) rod-line fishing with a nylon line and a steel hook, both of which are produced in urban areas, and a bamboo or wooden rod obtained in the local environment; (2) poison fishing, or fishing by using the poison of *Derris* shrubs (*D. elliptica* or *D. uliginosa*), which are usually raised in slash-and-burn gardens; and (3) shooting with bows and arrows of *waia.* Of these, rod-line fishing, which is a new method, is the most common at the present time. It is said that in olden times the people utilized a gorge-like device, made of the midrib of a wild palm frond, as a fishhook; a similar fishing device, which is made of fish bone, is reported for the oceanic peoples in the Pacific (e.g., Reinman 1967). About twenty kinds of freshwater fish (unidentified) including eel-like, catfish-like and carp-like fishes are caught using all techniques. The normal size of any variety is around 15 to 20 cm in length. Aside from fish, tortoises (perhaps *Emydura subglobosa*) were captured by rod-line, although I did not observe the catch.

Rod-line fishing can be done throughout the year by men, women, boys and girls. Actually, however, I did not observe *rugajog* or *nanyuruga* men do fishing by the rod-line method. *Kongajog* and *nanyukonga* women in groups and sometimes accompanied by *ngamugaibuga* girls, and *kewalbuga* males and *yambuga* boys in groups most frequently fished by this method. Also one *miid* man (Indiv. No. 1-1), who had retired from hunting, went rod-line fishing fairly frequently. The bait most commonly used are larvae of sago grubs (family Cerambycidae; perhaps *Rhynchophorus schach* Olive) found in the unworked stumps and husked outer skins of sago palms which had been worked for starch-making several days to several weeks previously. Alternately, the larvae of some insects inhabiting the inside of trees can be used for the same purpose. Everyone gathers twenty or thirty such baits on the way to the fishing ground. Most of the creeks and waterpools where the villagers fish are within an hour's walk from the village. At the fishing spot, each member of the group casts a baited line while crouching, sitting or standing.

Poison fishing and bow-and-arrow fishing are the exclusive practices of *kewalbuga* males, who are frequently accompanied by *yambuga* boys. Both methods can be utilized only in those waterpools which are formed through the drying-up of a creek in the mid-dry season. Usually, one or the other method is performed alternately along with hunting during the young people's expeditions. In fact, these methods yield very small catches, and the fish caught are usually cooked and eaten on the spot. Special fishing at the sacred place of Goara, which is done only by married and *kewalbuga* males, is very fruitful but is possible for only a few successive days in the dry season; a description of this type of fishing is reported in Appendix 4.

Collecting

Collecting occupies a minor place in Oriomo subsistence. The main products collected are plant foods, of which fruits of *Canarium vitiense* (galip nut) and *Gnetum gnemon* (jointfir, and called *tulip* in Pidgin) provide the greatest yield, while leaves of *Gnetum gnemon*, *Ormocarpum orientale* and *Abelmoschus manihot* (called *abika* in Pidgin) are next in importance. None of these plant foods are consumed at the spots where they are obtained, but are carried back to the village to be cooked. Except for *Abelmoschus manihot*, an annual shrub-like herb which is usually raised in gardens, none of the plants collected are cared for, nor are they owned by specific individuals or groups.

Canarium trees are fairly abundant in the area, and the villagers utilize those located within a half-hour's walk from the village. The tree is cut down from the bottom with an axe, and the fruits, each around ten g in weight and consisting of six g of meat and four g of inner shell, are plucked off the fallen tree. The reason for cutting the tree seems to be attributable to its abundance and its rapid growth, and, based on my observations, this tree species has never been exhausted around the village settlement. The yield from a single tree averages ten kg in weight, though there is great variation. The harvest season is restricted to the middle of the dry season, namely from September to November. A collecting party consists of one or two men belonging to *rugajog* or *kewalbuga* age-grades and several adult women and children. The men cut down the tree and the women and children collect the fruit. Since collecting parties usually include women and children from different households, the harvest is shared among several households. The fruits are boiled before being eaten.

Gnetum gnemon is cultivated in some areas of Papua New Guinea, for example in the Highlands (e.g., Clarke 1971: Appendix C); in my study area this species grows naturally, but the trees are few in number. The collecting of the fruit is usually performed in the early wet season, from December to February. Since only fallen fruit is collected, a gathering party is not organized. Instead, one or more villagers, mostly women and children, collect them, usually on the way back from the garden to the village. The collected fruit is not normally distributed since members of each household can collect the fruit as they like. A fruit is ten g or less in weight, of which 60% is meat and 40% is inner shell. Only the seed, weighing about three g, is eaten after being baked in heated ash.

Three kinds of leafy greens are available throughout the

year. Because Oriomo horticulture does not involve green vegetables, these leaves may be regarded as their substitutes, but the people do not frequently collect them, nor do they frequently plant the shrub-like *Abelmoschus* herb in the gardens. Women directly pluck off leaves from *Ormocarpum* trees two or three m tall, and gather the *Abelmoschus* herb, while it is usual for the *yambuga* boys accompanying the women to climb the tall *Gnetum* trees to collect leaves. The leaves are usually baked as a mixture with sago flour or steamed together with meat.

Other food plants collected include seeds of cycad (*Cycas circinalis* L.), fruits of a species of *Areca* palm (probably *A. macrocalyx*), a species of paudanus (probably *Pandanus brassii*), several kinds of palm shoots, and a great number of fruits of tree and grass species. Botanically-identified fruit-providing trees and grasses were *Aglaia* sp., *Beilschmiedia* sp., *Citrus aurantifolia*, *Decaspermum* sp., *Eugenia aquea*, *Finschia chloroxantha*, *Horsfieldia* spp., *Passiflora foetida*, *Pangium edule*, *Semecarpus* sp., several species of genus *Syzygium*, and *Terminalia ?microcarpa*, while the palms providing edible shoots were of seven or eight species belonging to genera *Areca*, *Arenga*, *Hydriastele*, and *Pinanga*. These are of less importance in terms of the amounts consumed, although they are edible without processing so that the people sometimes eat them in the forest and savanna while hunting or traveling for other purposes. These plants can be obtained by everyone, but in fact children of less than approximately ten years old do not usually collect them, depending on their own inclination.

In addition to the plants which are utilized by the villagers, I searched for natural growths of wild tuberous plants in the interests of ethnobotanical and prehistoric research, especially for a comparative study of the exploitation of wild food resources between Papuans and aboriginal Australians. I collected several specimens of four different local names, all belonging to genus *Dioscorea*, which were identified by botanical specialists as a single species of *D. pentaphylla* L. According to Powell (1976:178), this species is one of the yams which may be indigenous to New Guinea; or, in comparison with currently predominant species of *D. alata* and *D. esculenta*, it could have entered early from Southeast Asia or Indonesia (Burkill 1960); if so, *D. pentaphylla* yams would have been very important as an early food source in the savanna areas of the plain environment, including the Oriomo Plateau. However, the villagers say that *D. pentaphylla* can be planted in their slash-and-burn gardens, as in some other areas of New Guinea. Thus, there is still a possibility that such collected specimens are survivors of stocks whose ancestors once grew in abandoned gardens, although the origin of the specimens was lost to the villagers' memory. Along with the origin of the quasi-wild yams as well as the history of the yams predominantly cultivated in the area, any prehistoric significance remains to be found.

Another wild food plant which grows on both sides of the Torres Strait is the genus *Cycas*. The importance of this plant in the food register of the Australians in eastern Cape York Peninsula was emphasized by Harris (1977), who also suggested its importance among the Arnhem Land population. According to him, the main species of the Cape York Peninsula is *Cycas media*, while the Oriomo cycad was identified as *C. circinalis* L. The nutrient content of both cycad seeds is fairly similar, based on the analysis of *C. media* by Harris and *C. circinalis* in our 1981–82 study (Table 16). The nature of growth and the processing of the fruits, especially detoxification, are assumed to be identical in the two regions. The Wonie people collect the fruits, each containing five g of shell and six to seven g of kernel, in the early dry season and bury them under the ground, which is submerged with water in the wet season. After a year, the fruit is dug out and cooked on heated ash. Nonetheless, a distinct difference lies in the amount of consumption. The Oriomo Papuans utilize, at least at the present time, a much smaller amount of cycad fruit, although the density or the land productivity cannot be compared.

The animal foods which are collected include tree ants (*Oecophylla virescens*, and another *Oecophylla* sp.), sago grubs (family Cerambycidae; perhaps *Rhynchophorus schach* Olive),

Table 16. Food composition of cycad seeds (per 100 g)

	moisture (g)	energy (kcal)	protein (g)	carbo-hydrate (g)	fat (g)	fiber (g)	ash (g)	data source
Cycas media	49.91	201.28	5.09	43.52	0.22	0.55	0.71	Harris (1977)
Cycas circinalis	54.0	201	5.9	39.0	2.1	0.5	4.3	the present author

larvae of several species of insects which inhabit the inside of tree trunks, and honey. Of these, the most frequently captured is an assembly of *Oecophylla* imagoes together with eggs from a tree nest. However, all of these are very rarely gathered, and even the frequency of ant collecting is approximately several times a month by all the villagers, an assembly from a single nest being 100 to 200 g in weight. In some literature, especially Fountain's (1966) report of a study in the Sepik area, the importance of sago grubs as a protein source is emphasized. In such areas, men fell some sago palms deliberately to harvest sago grubs. Among the Oriomo, however, people collect small quantities of grubs from untreated portions of the worked sago trees, the stumps, the marginal inside of the trunk and the basements of fronds.

Exploitation of Sago

Sago palms growing in the area probably belong to only two species, *Metroxylon rumphii* Martius with a thorny foliar sheath and *Metroxylon sagu* Rottboell with a smooth foliar sheath, both of which provide starch for most sago-eating societies, while the genus *Metroxylon* in Oceania consists of eight or nine species, *M. amicarum* Beccari, *M. bougainvillense* Beccari, *M. warburgii* Beccari, *M. vitiense* Hooker, *M. upolense* Beccari, *M. salmonense* Beccari and presumably *M. oxybracteatum* Warburg ex K. Schumann and Lauterback as well as the above two (Barrau 1958:37, 1959). Sago palms grow ten to fifteen m in height and have trunks 50 to 75 cm in diameter. They yield palm flowers and fruit only once during a life span and continue to store starch in the trunk before flowering, thus reaching maximum storage just before flowering (Nishikawa, Oda and Suganuma 1979). After the flowering, the stored starch degenerates. For this reason the palm is not worth felling until it is at its vegetative stage or after it has passed the fruiting stage. According to Morris (1953:153–154), the maximum starch yield can be obtained in the early fruiting stage. It is generally stated in the literature that the maturation period is ten to fifteen years, but my own knowledge, based on this fieldwork, indicates a great variance in this period, depending on geographical and climatic conditions; palms in permanent swamps attain maturity by ten years or

Table 17. Plant foods raised by the people

common name	local name	scientific name (genus)	no. of local names	no. of those names affirmed[a]
sago	sana	*Metroxylon*	23	14
coconut	gujo	*Cocos*	11	4
garden crops[b]				
taro	wor	*Colocasia*	62	25
	kerema	*Alocasia*	1	1
banana	sela, kemba	*Musa*	41	16
papaya	lameap	*Carica*	2	2
pineapple	paniap	*Ananas*	4	4
elephant-foot yam	geb	*Amorphophallus*	5	3
yam	sia, sula	*Dioscorea*	22	3
cassava	poani	*Manihot*	6	2
sugar cane	gau	*Saccharum*	12	6
sweet potato	nger	*Ipomea*	8	5
pandanus	gereba	*Pandanus*	5	4
pitpit (in Pidgin)	usu	*Saccharum*	1	1
aibika (in Pidgin)	beag	*Abelmoschus*	1	1

[a] This category includes the locally-distinguished kinds which two or three aged men of Wonie affirmed through identifying the plants themselves.

[b] Excluded from this list are pumpkin, watermelon, cowpea and breadfruit, for reasons mentioned in the text.

less, while palms in seasonally-submerged environments attain maturity after some twenty years, especially if the palms encounter any drought early in their growing stage. After the palm tree bears fruit it gives out a number of shoots from its root system, and, therefore, sago stands tend to be naturally dense.

The two species (*M. rumphii* and *M. sagu*) have more than ten names for the locally-recognized kinds (Table 17). However, all kinds are grown by the people, without specific distinctions as to use, and the size and yield per tree are not distinguished among them, according to my observations. Sago exploitation involves only the initial planting of a shoot, although even this is not necessarily required, and the eventual harvesting, i.e., starch-making; harvesting follows planting by ten to fifteen years. Both planting and harvesting can be done throughout the year, irrespective of the seasonal climatic cycle. The transplanting of dissected shoots is done by the married men who will be the owners of the respective palms.

This procedure involves only digging a hole for each shoot and putting the shoot in it. In addition, some of the utilized sago stands have been self-propagated. After planting, no special care is required until harvesting, but a married man will inspect his own sago palms while on hunting expeditions or while the women of his group are actually making sago at a grove. These frequent visits play a useful role in determining the right time to cut palms. The visible signs useful for obtaining the maximum starch yield are not certain, despite the fact that the villagers know that the palms with flowers give a large amount of starch. This is because the extreme density of the palms usually makes it impossible to see the flowers. The Wonie villagers say that some harvested sago trunks are not full of starch when felled and a few do not contain starch at all, although I failed to ascertain the exact proportion of such cases. I observed eight sago-makings and found a great variation in starch yield per tree, from 29 to 104 kg. This was, perhaps, mostly attributable to the different times when cutting took place.

Sago-making is carried out by groups which are informally and temporarily organized. The sago-making parties usually include not only the owner of the palm to be cut and his wife or wives but also, in many cases, other villagers who wish to obtain starch at the moment. Brothers and sisters of the owner, together with their respective spouses, may join the sago-making parties. Also, a young married couple who possess a small number of matured stands as well as wives of polygynous families will cooperate with close relatives such as parents and brothers. Moreover, a widow may join any group.

It is the men's task to cut down the sago palm, about 40 to 70 cm above the ground, with an iron axe. When the palm is lying on the ground, they remove the upper half of the surface wood with the axe. The thickness of this wood is around two to four cm along the length of the tree, and its removal exposes the pith which fills the whole interior of the trunk (Photos 13 to 17).

The women then begin scraping out the sago pith and pounding it into a fibrous mass. This work usually involves two or more women, each performing the same task at separate points along a single trunk. The pounding is done with a special tool called *abol* (Photo 15), which is made of two hard wooden sticks and a tough string of cane (*Calamus* sp.). A stick for the handle shaft and another stick with a concave-shaped striking surface are kept in the village house for repeated use. This implement is operated in the same way as an adze and at 1.0- to 1.2-sec intervals on the average. When the loosened pith is accumulated, the second stage of the operation, washing out the sago starch from the pith, commences. This task takes place in a sort of trough about 1.8 m long, which is made of a sago frond. The wider end is raised about 60 cm to a slanting position with a strut made of either a sago frond or wooden sticks. Fixed just above the lower end of the trough is a filter made of a spathe of coconut palm. The woman engaged in washing out the sago fills the trough with two handfuls of the pounded pith and pours in a large amount of water, which she brings from a small water hole near the working place. Then she begins kneading the pith with both hands, pouring more water in at intervals. The number of kneadings for a lump of pith averages four or five. The suspended sago starch runs out with the water through the filter into a vessel made of a dried frond of wild palm, while the pith remains in the trough. In this fashion the vessel is filled with water, dark orange or brown in color, and the starch is left to sink gradually to the bottom. After a while, the woman removes the water by tilting the vessel, leaving a white powdery sediment which dries into lumpy sago flour.

Case record of a practice of sago-making.

1) Date: 20 January 1972.
2) Location: Six km west of the village, called Wurbit locally.
3) Owner of sago palm utilized: Indiv. No. 1-1.
4) Local name of the palm utilized: *wurmopgua* belonging to *Metroxylon sagu.*
5) Participants (Indiv. No. and age-grade):
   1-1 (*miid*), 1-2 (*nanyukonga*) 13-1 (*rugajog*), 13-2 (*kongajog*), 4-3 (*kongajog*), 17-2 (*kongajog*), 19-4 (*nanyukonga*), K-9 (*kewalbuga*). In this sago-making party, the proportion of women who were not accompanied by their husbands is higher than usual. The *kewalbuga* (K-9) is a son of the owner.
6) Labor time:

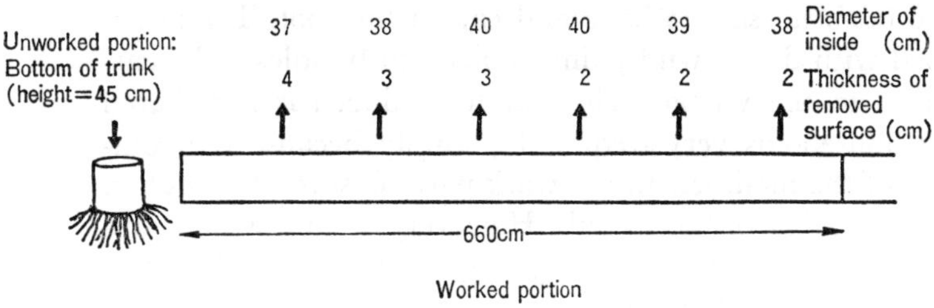

Fig. 14. Worked Portion of Sago Palm.
The bottom of trunk, which was the unworked portion, is illustrated on the left.

Between 8:50–10:05. The palm was cut down and the surface wood was removed, mainly by Indiv. Nos. 13–1 and K-9. During this time, five women individually made troughs for washing out sago starch within several meters of the palm.

Between 10:03–10:17. Five women started pounding out pith, each with an *abol* at separate points.

Between 10:25–10:48. Five women started washing out pith. After that, two different tasks were continued repeatedly until approximately 17:30. According to my estimation, about 70 percent of the time between 10:00 and 17:30 was used for actual labor, but without undue exertion. As the women were working, Indiv. Nos. 1–1 (a *miid* man) and K-9 (a *kewalbuga* male) left the working place for the village at 13:30, while Indiv. No. 13–1 (a *rugajog* man) stayed on except for a couple of hours' expeditionary tour with a bow and arrows in his hands.

7) The worked portion of sago palm is shown in a simplified illustration in Figure 14.
8) Amount of crude sago produced by the individual women:

Indiv. No.	amount of sago (kg)
1–2	8.6
4–3	12.1
13–2	13.0
17–2	14.5
19–4	6.5
total	54.7

When the sago-making is over, the women usually wrap a small part of the sago in young leaflets of sago palm or leaves of cane (*Calamus* sp.) to bake and eat on the spot. The rest is wrapped with dried, wild palm fronds into bundles to be carried back to the village. The starch produced is not shared among households very frequently, simply because the composition of the members in the work party ensures distribution among a number of households. However, when a woman has been sick for a long time or has just given birth, starch is usually donated to her household.

Planted stands and self-propagated stands of sago have not been distinguished here because the villagers themselves do not care whether the sago to be utilized has been planted or is self-propagated and because my data suggest that there is no significant difference in size or in starch content between the two types of sago palm. Such a conclusion is supported, at least tacitly, by other workers (e.g., Beaver 1920:161; Conroy and Bridgland 1947:73; Barrau 1958:38), who describe the coexistence of wild stands and planted stands in their respective study areas without drawing any clear distinctions between them. As shown in Table 18, however, starch yield differs among the sago-exploiting populations: P. K. Townsend (1974) reports a starch yield of only 0.08 kg/m^3 from wild palms of the Sanio-Hiowe inhabiting the Sepik region in comparison with 0.2 kg/m^3 from planted palms of the Abelam also inhabiting the Sepik region (Lea

Table 18. Crude starch yield per tree of sago palm among the Oriomo and some comparable data from other societies

place	number of trees studied	weight of crude starch (kg): average (range)	property	data source
Oriomo	8	66 (29–104)	wild & planted	the present author
Kiwai I., estuary of the Fly	—	90–140	wild & planted	Beaver (1920)
Abelam, east Sepik	2	219	planted	Lea (1964)
Sanio-Hiowe, upper Sepik	5[a]	86 (28–205)	wild	Townsend, P. K. (1974)
Irian Jaya in general	—	110–160 (highest=410[b])	planted	Barrau (1959)
Baroi, Purari delta	18	134 (38–359)	planted	Ulijaszek and Poraituk (1981)

[a] Among the seven trees reported, two are omitted here by the present author, because the sago-working of the two trees was not fully observed.

[b] Naturally sterile palm.

1964) and concludes that the Abelam may have better-yielding sago because only selected varieties are planted. A more practical measure for evaluating the importance of sago exploitation as a subsistence strategy may be labor productivity. According to Townsend, 154 man-hours are required to obtain one million kcal for Abelam sago and 157 man-hours for Sanio-Hiowe sago. Comparable data from the Oriomo indicate 180 man-hours (when applying Peters's (1957) food composition table indicating 285 kcal of energy for 100 g of raw sago, as in Townsend's study, although this value of energy content will be discussed later in terms of nutrient analysis in the present study). Judging from these figures, it appears that similar amounts of starch per unit of labor are produced, irrespective of the wild or planted status of the sago palm.

Techniques of sago-making involving felling palms, pounding pith and extracting starch, which the Oriomo Papuans use, are similar throughout New Guinea, although variations exist. For example, the Keraki of the Morehead River and the Wiram of the middle Fly do not wash out pounded pith in a trough, but pack it into a close-woven bark bag and then stamp on it (Williams 1936:422–423). In terms of the sexual division of labor, pounding and washing are done mainly by men among some other peoples, e.g., the Dobuans (Malinowski 1922:377–378), the Wogeo (Hogbin 1938–39:306–307) and the Tikopians (Firth 1939:135). However, these groups depend on sago much less than the Oriomo Papuans.

Exploitation of Coconut

Coconut palms consist of four locally-distinguished kinds (Table 17), all of which are exploited by the people without any distinction as to usage. The ownership, inheritance, planting method, year-round availability and organization of harvesting parties (and thus the distribution of the products) in coconut exploitation and sago starch exploitation are almost identical. This is because coconut palms, like sago palms, take ten to fifteen years to attain maturity. In the organization of a harvesting party, however, there is a minor difference in that the aged men belonging to the *miid* and *nanyuruga* agegrades rarely take part in coconut-collecting parties because they are unable to climb up the palm to fell the fruit. Nonetheless, the exploitation of coconuts differs from that of sago in three important ways: first, coconuts are much less important as food than sago starch; second, coconut palms can grow in a much wider area of savanna and savanna woodland in the village-land, although because of the smaller demand a limited part of such lands have been actually exploited; third, the harvesting of coconuts is much less complex and time-consuming, that is, harvesting coconuts merely requires that the men climb the trees to obtain the fruits and that the women husk them.

Horticulture

Plants reared in slash-and-burn gardens are rich in variety, including twelve kinds of food plants as well as other useful plants (several species of bamboos and/or bamboo grasses, i.e., *Bambusa vulgaris*, *B. atra*, etc., for material for bows and arrows, and poisonous shrubs, i.e., *Derris elliptica* and *D. uliginosa*, used in fishing), although the non-food plants are perennials and consequently when they are ready to be utilized all food plants in the same garden have normally been harvested out. Like the perennial non-food plants, several trees of breadfruit (*Artocarpus altilis*) grow in old gardens throughout the village-land. Breadfruit was introduced to this region, possibly at a later stage than other crops, but is not utilized as an important dietary source. The growing and harvesting of breadfruit trees is the same as for the naturally occurring species of *Gnetum gnemon*.

As shown in Table 17, there are twelve garden crops, although excluded are pumpkin (called *ponke* locally), watermelon (*wayati*) and cowpea (*tamatama*), which have just been introduced to Wonie and are grown, very small in size, by a few households. Of the twelve crops, the fruit of pandanus *Pandanus ?conoideus*) is used as a kind of seasoning. *Amorphophallus campanulatus* Bl., sometimes called elephant (-foot) yam, is one of the garden crops, and the tuber is eaten by the people as same as taro and yam, either baked on a fire or steamed as a mixture of meat; nonetheless, according to Purseglove (1978:59), the tubers contain needle-crystals of calcium oxalate, which must be removed by prolonged washing or boiling. Massal and Barrau (1956) pointed out that *Amor-*

phophallus tuber was perhaps formerly of much greater importance as a subsistence crop in the Pacific than at the present time. In spite of this fact, Powell (1976) in his review on ethnobotany in Papua New Guinea lists this plant only as a medicinal plant. The Wonie people recognize more than thirty species of medicinal plants, most of which grow naturally, but *Amorphophallus* is not included in this category.

The number of local names for the kinds of the particular food plant, which were not and perhaps cannot be botanically distinguished from each other, is listed in Table 17, where, exactly speaking, the total number of local names collected and the number of those names which were affirmed by two or three aged men of Wonie are shown. The various names are based not only on the shapes and colors of leaves, stalks, fruit and/or tubers (or their specific parts) but also on historical knowledge such as the believed place of origin. In fact, the people identify these locally-recognized kinds fairly well based on their growing state in the garden. The plants with edible fruit were identifiable as to kinds to some extent, while I found the identification of the kinds of tuberous crops almost impossible. Roughly speaking, the more the names for a single food plant, the more ancient and the more important it is as a food source. The two most abundant crops, taro and banana, have a tremendous number of names, while taro of the genus *Alocasia* still possesses only one local name at the present time. The *Alocasia* taro was recently introduced, within the memory of the present villagers, and provides a very small yield. Most kinds of banana are plantains, although some of them can be eaten raw after full maturation. Except for *Alocasia* taro and bananas, there are no remarkable differences among the locally-distinguished kinds of each food plant in regard to shape and size of the edible portion, seasonal availability or method of processing.

Not all of the garden crops can be harvested throughout the year. Of the major crops, taro (both *Colocasia* and *Alocasia*), yam, sweet potato and *Amorphophallus* can be harvested only in the dry season; sugar cane and *pitpit* (in Pidgin) mainly in the wet season; while banana, papaya, pineapple, cassava and *Abelmoschus* leaves are harvested at any time. Although it is difficult to distinguish season because of the small harvests, pandanus may be harvested irrespective of the seasonal cycle.

Furthermore, the harvesting season of bananas may differ from kind to kind, although I could not determine this fully.

Due to the seasonal availability of some crops, especially taro, horticulture requires a yearly cycle of labor. In the early dry season, from August to October, members of each household clear one or two large new gardens, 150 to 200 square m in area on the average. The men fell trees with axes while the women clear shrubs and grass with bush-knives. When dry, the grass, shrubs and trees are burned in middle or late dry season, mainly in October and November (Photo 18). After this, planting takes place. Before and/or after the planting, however, the gardens, especially those located far from the village, are occasionally fenced in by men to prevent damage by wild pigs. Apart from clearing the intended garden plots and fence-making, most horticultural labor is performed mainly by women. The planted parts of the major food plants are cuttings, shoots or tubers (Table 19), which are usually fetched from the old gardens just prior to transplanting, although yam tubers are frequently stored inside the dwelling house or in other dark places to produce buds to be planted. No specific nursery is made for any kind of crop. The planting is done in many cases by women using digging sticks, which are not elaborately shaped but which are made of nearby tree trunks for temporary use. The procedure of planting is simple, consisting of digging a hole and putting each cutting or shoot and then a small amount of mud in it (Photo 19). After the planting of taro and some other crops in the early dry season,

Table 19. Planted parts of garden crops

crop	planted part
taro	cutting from the top of the corm or the axillary bud (shoot)
banana	shoot
papaya	stem cutting
pineapple	cutting from the top of the fruit or shoot
Amorphophallus	small tuber
yam	cutting from the top of the corm or the axillary bud (shoot)
cassava	stem cutting
sugar cane	shoot
sweet potato	leafy shoot
pandanus	? shoot
Abelmoschus	stem cutting
pitpit (in Pidgin)	shoot

a new garden is unsystematically weeded once or twice, generally by women. Old gardens where only perennials grow are never weeded.

The gardens are owned by a household. However, each garden plot of a polygynous household is partitioned according to the number of wives, one of whom utilizes her portion with the husband, although the husband has responsibility for the male labor in all the portions and has latent co-ownership.

An example of a large new slash-and-burn garden.

1) Owners: Household No. 2 and Household No. 18, both of which cooperated in making the garden. The wife of the former (Indiv. No. 2-2) and the head of the latter (Indiv. No. 18-1) are sister and brother.
2) Location: Five hundred m northeast of the village, called Kemop locally.
3) Time of observation: 5 February 1972. The planting of taro was almost completed, but banana, papaya and some other crops were planted little by little after the observation date.
4) A sketch map of the garden is shown in Figure 15.
5) Number of plants grown at the time of observation:

    In the garden of Household No. 2:
    taro—ca. 600
    banana—ca. 80

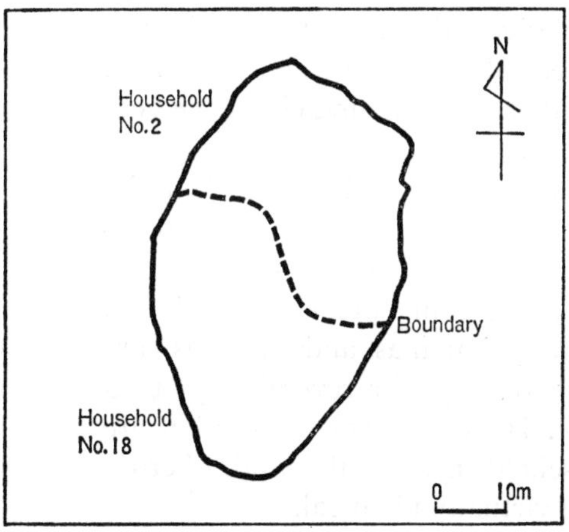

Fig. 15. Sketch Map of New Gardens.
The boundary between the gardens is marked by several fallen trees.

pineapple—3
yam—8
In the garden of Household No. 18:
taro—ca. 450
banana—ca. 120
yam—3
cassava—5

In addition to large gardens, every household annually makes several small gardens adjacent to the village or at the margins of old gardens. Making these small gardens does not involve clearance of trees and shrubs, and the season for this work is not fixed. Except for some of these small gardens, in which only ten or twenty cuttings and/or shoots of one or two kinds of crops are planted, the kinds of crops raised in both large gardens and small gardens are almost identical, the proportion of taro being much lower in the latter. The usual combination consists of taros, bananas and a small amount of one or several other crops. Due to the intermittent planting of bananas and certain other crops, however, the crops in a garden tend to change from time to time. While annual plants will be completely harvested out in a year, perennials such as bananas and papayas are left growing, and thus the gardens, especially the large ones, will be utilized for several years (Photo 20). For this reason, the number of gardens possessed by each household ranges from five to twenty at any given time; all the gardens of each household used from December 1971 to February 1972 are listed in Appendix 6, together with the first clearance year and the kinds of plants involved.

Harvesting requires little labor input and is done little by little as food is required. Normally a member of each household, usually the wife, fetches various kinds of crops almost every day. Most crops are eaten immediately or within a week, although fruits, especially bananas and papayas, are normally harvested unripe to be kept for several days to a couple of weeks in the house. Harvests from the garden are rarely shared among the households because the kinds of crops raised by every household are virtually identical.

2. Foodstuffs and Utilization

Availability and Nutrient Composition

Since no cash crops are produced and no purchased foods make any significant contribution to the diet of the area, the products of the subsistence activities mentioned above comprise the menu of the Wonie villagers. Judging from the daily consumption, the number of foodstuffs forming their basic menu is limited. Animal foods, which are all *ferae naturae*, are obtained almost entirely by hunting, and three species of wallabies, pigs, cassowaries and two species of bandicoots occupy the bulk of it. Subsistence activities other than hunting provide the people with a very small amount of animal food, largely freshwater fish and tree ants with eggs. Sago, banana and taro comprise the bulk of the plant foods, while the secondary important foods obtained by horticulture are papaya, pineapple, yam, sweet potato and *Amorphophallus*, and those obtained by collecting are only the fruits of *Canarium vitiense* and *Gnetum gnemon*. Coconut fruit is involved in the latter category.

Although no food samples were collected in my own 1971–72 survey for the laboratory analysis of the nutrient composition, our group fieldwork in the new project of 1980 and 1981–82 involved sampling foodstuffs which were major diet sources or whose nutrient composition was unknown, and analyzing their major nutrient and metal concentrations. The foodstuffs studied numbered 82. The results of this study are beyond the scope of the present volume, but the basic information concerning the major nutrients and the energy involved in the foodstuffs (Table 20) will be applied to the food which appeared in the 1971–72 food consumption study. In comparison with several published food composition tables, a remarkable difference lies in the greater amount of water content in our specimens, especially sago flour, largely because our specimens were initially weighed just after production or at the time of consumption in the village, while the foodstuffs examined in the other studies might have been obtained from local markets away from the production areas. In this respect,

Table 20. Food composition table of major foods (per 100 g edible portion)

common name	scientific name	no. of samples	moisture (g)	protein (g)	fat (g)	carbohydrate (g)	energy (kcal)
sago	*Metroxylon* spp.	2	44.5	0.4	Tr.	54.8	220
coconut: meat	*Cocos nucifera*	A	36.0	4.7	53.4	9.9	479
coconut: juice		A	92.7	0.4	1.5	4.6	32
banana	*Musa* spp.	8	73.8	1.1	1.1	26.1	119
taro	*Colocasia* sp.	3	66.6	1.8	0.1	30.6	130
papaya	*Carica papaya*	A	87.0	0.5	Tr.	12.1	45
pineapple	*Ananas* sp.	B	84.3	0.5	Tr.	11.6	46
yam	*Dioscorea* sp.	2	76.1	1.5	0.1	21.3	92
sweet potato	*Ipomea batatas*	1	80.6	1.1	0.7	16.6	77
galip fruit	*Canarium vitiense*	1	59.2	2.0	14.5	22.0	226
tulip fruit	*Gnetum gnemon*	1	73.0	2.9	0.4	22.9	107
meat		3[a]	73.4	21.5	0.7	0.3	94
fish		5[b]	80.1	15.2	1.9	1.5	84

Figures for the nine food items are based on the analysis during our 1980 and 1981–82 study; when two or more samples were analyzed the figure shows the mean value, although (A) those for coconut and papaya follow Hodges, Fysh and Rienits's (1947) food composition table, and (B) those for pineapple follow McCance and Widdowson's (1960).

[a] Mean value of three samples of grass-wallaby, bush-wallaby and scrub-wallaby.

[b] Mean value of five different freshwater fish species.

Tr.: traces of constituent are known to be present, but in any case the amount is of no quantitative dietetic significance.

Table 21. Changing weight of sago flour through drying up (percent of the initial weight)

no. of days from the production	0	5	10	15	20	25	30
Sample 1	100	87.5	82.5	80.0	78.8	77.5	76.3
Sample 2	100	89.2	85.0	82.6	79.9	74.4	

Both samples at the final measurements were so dry and crumbly that the villagers did not want to eat them.

Oomen (1971) points out that raw sago contains from 20 to 45 percent water. I examined the changing weight of sago flour in the course of drying (Table 21) and found that the reduction in water content was one-fourth over a 25- or 30-day period.

In terms of the nutrients involved, all the major foodstuffs are categorized as follows: (1) sago flour, (2) coconut, its meat and juice showing of course very different values, (3) banana and tuberous crops (taro, yam, *Amorphophallus*, sweet potato), (4) papaya and pineapple, (5) collected fruits, (6) land mammals and cassowary, and (7) freshwater fish, of which all animal foods (Category Nos. 6 and 7 above) are characterized by a similar food composition. In spite of a fairly large variation in nutrients among bananas and tubers in Category No. 3, an overall comparison leads to their classification within the same group, and the villagers themselves recognize them as similar food items. The above categorization corresponds to the variety of subsistence activities, with only Category Nos. 3 and 4 being produced by horticulture.

Apart from the chemical composition of foods, the villagers' sense of taste distinguishes minute differences among foodstuffs. This ability includes, for example, not only the inter-species and inter-(locally distinguished) kind differences among tuberous crops and those among animals, but also differences in sago flour according to the upper or lower portion of the tree from which it was extracted and the duration after production. The villagers are aware of the frequency of intake of food items and of the combination of food items eaten at the same meal, although the availability of the various foods determines these matters to a great extent. Each meal involves two or three food items on the average, and three meals are usually taken in a day. Continuous intake of a single foodstuff for four or five meals is avoided as far as possible. In terms of the combination of foodstuffs, the villagers prefer to eat plant staples, sago, some tuber or banana, and meat together, with "delicious" foods like pineapple, papaya and tree fruits being added if available.

Due to the hot, humid environment in the area, the storage time for all foodstuffs is fairly limited. Sago flour is usually consumed within a couple of weeks, although it is commonly believed that sago may be stored easily for many months (e.g., Ruddle, Johnson, Townsend and Rees 1978:27). The short storage time among the Oriomo Papuans may be related not to edibility but largely to taste and to frequent sago-making. The garden crops are consumed within a week or so after harvesting, except for the maturing time required for cropped plants (especially such fruits as banana and papaya) in the dwelling house. It is also noted that the edibility of tuberous crops is much longer than they are actually kept in my study village, although I failed to examine the potentially maximum length. Regarding animal foods, small- and middle-sized animals about five kg or less, such as bush-wallabies, are largely consumed within a couple of days. The meat of large game animals tends to be consumed successively; that is, the people eat the surface portion of the baked meat and then bake the rest again to repeat the same process. Even in such cases, however, the meat lasts for only a week at most. The cassowary is the only game animal that has a thick layer of fat (about two cm thick), and this fat need not be consumed for a week. The actual storage time of animal meats also depends largely on taste and the villagers' preference for consuming any food on hand within a short duration; this is manifestly supported by the unpredictable nature of catching game animals, and by the presupposition of a fairly large supply in the long-run.

Food Processing and Food Habits

Of plant foods, coconut, papaya, pineapple, sugar cane, some kinds of banana other than plantains, and several species of collected wild plants are eaten raw, while the rest require some processing. The cooking method is almost exclusively baking because, in spite of the introduction of steel pot cooking some decades ago, the lack of traditional pottery in the area makes boiling uncommon. Sago flour, taro and many kinds of banana (plantains), which constitute the staple food group, are normally cooked on a fire. Sago is wrapped with sago pinnae or cane leaves while the others are cooked on a direct flame. Another traditional cooking method of the major plant foods is to steam them together with animal meat, as mentioned later. Plant foods are, as a rule, not distributed among households, except for those who participate in sago-making, coconut-fetching and *Canarium*-collecting parties, which enable the participants to obtain products of their own. This lack of food distribution is simply because all household members can easily obtain identical foods.

All animal foods, with such minor exceptions as honey, necessitate cooking. Fish and sparrow-sized birds are put directly on the fire, like taro and banana. Pigeon-sized birds and terrestrial animals less than several kg in weight are cooked in a similar way, although prior to the baking, feathers are plucked off for the former and the body is dissected for the latter. The cooking of these smaller animals is done by an individual or members of a household, and the cooked foods are consumed mostly within the household, while relatively large animals like bush-wallabies are presented in portions to several households. This depends upon the judgment of the members of the household concerned.

The cooking of large animals, cassowaries, pigs and full-grown grass-wallabies, is performed in a kind of "casserole" by many villagers through a complex method (Photos 21–24). The base of the casserole is a sort of scaffold with several stairs of logs, about 50 to 80 cm in total height. Each stair is composed of several logs about one m long which are laid out parallel with spaces between each log. The logs of the stairs are arranged alternately in different directions. Small pisolitic laterites or broken pieces of termite hill, which are stored by most households, are heaped on this structure. Pisolitic laterites are fairly common in the Morehead Ridge, the relatively higher part of the Oriomo Plateau which includes the study region, and termite mounds of, perhaps, *Nasutitermes triodiae* form one of the characteristic features of the Oriomo Plateau (Bleeker 1971). A fire is set under the logs, and, after a while, the pisolitic laterites or termitaria become intensely hot. Concurrently, other cooks dissect the animal meat into several large pieces, keeping the bones on, and roast them on the burning fire for a minute or so to take away feathers or fur. They put the entrails and small pieces of meat in containers made of a frond of wild palm, mixed together with sago flour, bananas and/or other plant foods, as well as water and, in some cases, a small amount of the animal's blood. The large pieces of meat with bone and the packed containers are placed on the fully-heated pisolitic laterites or termitaria when half-burned logs have been removed. This casserole is then covered with green banana leaves and the bark of the *ti* tree (*Melaleuca* spp.). The final stage in this cooking process is to steam the food under cover for an hour or an hour and a half.

This communal cooking involves several adult men and women who are not organized on any special basis, but whose involvement in the cooking is fairly arbitrary. It is usual that the hunter who shot the animal to be cooked does not participate in the cooking. In this system, the hunter in question is not awarded a larger share, nor specific portions. The generous and modest behavior of the hunter is greatly honored, however. This cooking ends in an impartial distribution of the animal meat among all households, roughly according to the number of members, with certain exceptions because of the food tabu systems, as explained below. All animals obtained by group hunting are impartially distributed to all households, and the same cooking system is applied. During my study period, more than half of the animals consumed among the villagers were prepared by this cooking system. In addition, the smaller animals caught by individual hunters were also delivered fairly equally to all households over a long-term period. In conclusion, the sharing system ensures the impartial consumption of animal meat among all the villagers.

There is no traditional prohibition against eating any animals or plants, nor their specific portions. This applies to all Gidra-speaking members. Apart from feathers, fur, bones, substances in the entrails, and the gall bladder, the animal is completely eaten, although the bone marrow is not always eaten. The major reason why edible plants which are easily obtained are not always eaten is usually attributable to their unpalatable taste, although all such plants are wild and are not collected unless the people intend to eat them. Nonetheless, the people have a food tabu system which is related to clan organization. There is a prohibition against eating *kakiena*, the principal totemic animal of the clan, as well as against capturing it. Nowadays, however, this system is not strictly observed, and only the aged villagers maintained this tradition during my study period. These aged villagers were assured amounts of animals other than their *kakiena* so their total meat consumption was almost equal to the average level for the village.

Aside from this, two other food tabus are applied to people who are in critical life stages as determined by tradition. The first system, still observed to the present day, involves *kewalbuga* males. According to the original rule, when a boy enters the *kewalbuga* age-grade, the foodstuffs available to him include only sago, taro, banana and pineapple. After that, as married men gradually give him the foodstuffs which have been prohibited, his menu increases in variety. The major foods which are subject to lengthy prohibition include wallabies (grass-wallaby, bush-wallaby and scrub-wallaby), bandicoots, pigs, cassowaries, and all snakes and lizards, in increasing order of length of prohibition. On the other hand, all plant foods, fish and small birds are eaten freely after several weeks of *kewalbuga* life.

During my study period, a male (Indiv. No. K-9) entered the *kewalbuga* age-grade in September 1971. Since the new *kewalbuga* spent much time outside of the village settlement, direct observation of his diet was almost impossible. Information based on interview and limited observations made clear that at least for the first week he ate only the four kinds of plant staples mentioned above, but after that his menu items rapidly increased; meat of wallabies, which are the most readily available larger animal, was given to him ten days after the change of age-grades. Another *kewalbuga* male (Indiv. No. K-8), who was the second youngest *kewalbuga* and had entered this age-grade about two years earlier, told me that he was still forbidden to eat cassowary, snakes and lizards, including monitor lizards. The other *kewalbuga* males, who were more than two or three years senior, could eat cassowary, so the animals which were subject to several years' prohibition were only snakes and lizards, both of which were rarely killed.

The second type of food tabu system applies to women who have just given birth. However, the food prohibition in this system appears to be in effect, at least at the present time, for a very short period of several days, and there may be no practical influence on food consumption, although I failed to obtain accurate information. Considering the current prohibitions both for new mothers and for new *kewalbuga*, the food tabus do not profoundly affect the people's nutritional state.

The basic meal unit is a household, while a polygynous household is divided into different units, the number of which equals the number of wives, and the husband usually belongs to one of them. The formation of meal units parallels the ownership of horticultural gardens. On the other hand, *kewalbuga* do not belong to any family unit, but take meals individually or in groups of age-mates. The bulk of the *kewalbuga*'s diet is given to them by various married people, and the foods presented have been usually cooked. *Kewalbuga* frequently snatch cooked food from married people during eating hours.

The frequency of meals in a day is not fixed, although villagers normally eat staple foods such as sago, taro and banana three times per day, ranging from two to four times, and, if snack-like food consumption is taken into account, their meal times may exceed four. The actual time of the meals and the intervals between them vary depending on the conditions of the day in question. Relatively fixed meal hours are in the morning between 7:00 and 10:00 and in the evening at or after sunset, while other meal hours are very irregular.

4. Subsistence Adaptation

1. Subsistence System

The foregoing chapters have given descriptions of the people of Wonie, of the village-land as their focus of activity and of the food-getting and -consuming activities of the villagers. The present chapter is concerned with the subsistence system which quantitatively relates the subsistence activities to the people on the one hand and to the village-land on the other. In other words, this chapter concentrates mainly on the elucidation of the structural aspects of the villagers' subsistence. Quantitative data and analytic discussion will be arranged on the basis of the five following categories: (1) food consumption, (2) time allocation, (3) labor efficiency, (4) labor units and (5) use of space. Food consumption, the first area, is concerned with the results of their food-getting activities. Time use (time allocation) and space use (use of space) are regarded as the basic components of a human activity system, although two subjects, i.e., labor efficiency and labor units, are inserted in between because these two are closely related to time use.

The systematic and quantitative data for all of the categories to be discussed were obtained by three types of investigation. One type treated food consumption and the other two treated food-getting activities. Food consumption was studied in the following manner. Due to the irregular daily meal hours of the people and the difficulty of applying a questionnaire survey as in usual food consumption studies for modernized societies, data are corroborated by the precise measurement of all foodstuffs carried intermittently to the village, as in Rappaport's (1968) study of New Guinea High-

Table 22. Family composition of households for food-consumption survey

Household No.	head	housewife	children
1	1 (*miid*)	1 (*nanyukonga*)	2 (1 *yambuga*; 1 *sobijogbuga*)
6–1[a]	1 (*miid*)	1 (*nanyukonga*)	2 (1 *ngamugaibuga*; 1 *sobijogngamugai*)
6–2[a]	0	1 (*nanyukonga*)	3 (1 *ngamugaibuga*; 2 *sobijogngamugai*)
14	1 (*rugajog*)	1 (*kongajog*)	2 (2 *sobijogngamugai*)
19	0	1 (*nanyukonga*)	0

[a] Household 6, which was polygynous, was divided into two meal units.

landers. In this framework, it was also indispensable to take into account some other types of food flows which occasionally occurred. For this, I measured all foodstuffs which, either raw or cooked, were transferred between different households or from married people to *kewalbuga* males in the village settlement and which were occasionally carried out to sago groves or gardens to be consumed. Due to these conditions, data gathering necessitated constant observation, and therefore the number of subjects for the study was limited. The survey unit for food consumption was a household which formed a meal group. Four households were selected (Nos. 1, 6, 14 and 19) consisting of 17 persons, three married men, five married women and nine unmarried children, excluding school pupils who were out of the village throughout the investigation periods. As shown in Table 22, these four households varied in organization. They included a widow's household and a polygynous household which was divided into two meal groups.

Since their food consumption fluctuates from day to day and varies throughout the year, especially between the dry and wet seasons, it is necessary to carry out the investigations of the two seasons for successive days. In fact, surveys of twelve successive days were conducted twice for the four sampled households, once in the dry season and once in the wet season. In each survey period, all foods which were stored in the respective houses at the beginning and at the end of the period were measured, in addition to measurements of all food flows which occurred during the period. In order to assure accurate data, the food consumption records for the days when the people stayed overnight out of the village were excluded from the analysis. This reduced the net survey days; the total number of [meal unit]×[day] that could be studied was 60, but the actual total numbers of [meal unit]×[day] used for the analysis were 46 and 48 for the dry and wet seasons, respectively. Another procedure was to inquire frequently about whether while outside the village they had eaten some foods which had not been carried out from the village on the net survey days, although such cases occurred very seldom. In these cases, I weighed similar substitutes of the same species.

Food-getting activities were recorded by two types of survey. The first was an intensive time-recording survey, in which I recorded the time of departure from and return to the village from morning till evening for all adult Ginar-dwelling villagers, checked the accompanying people and baggage, and asked them upon their return where they had been and what they had done. In this investigation, I was helped by my assistant who simultaneously watched the people's movement at the opposite end of the village because the Wonie villagers had to pass through either of the two ends, from which the paths extended. Noting seasonal variations and daily fluctuations, the time-recording survey, like the food consumption survey, was carried out twice, in both the dry and wet seasons, for thirteen successive days each.

The second type of survey was an activity-tracing survey, in which I followed the people, mainly Ginar-dwelling adults, to their working places to observe and describe their activities in as much detail as possible. In this study, I greatly feared that my participation would disturb the villagers' actions when their work was less than totally demanding, as is typical in horticultural work. This might have been the case in some instances, but the information obtained in this study was indispensable to my detailed understanding of food-getting activities, and only the data that were gathered under conditions of minimal or negligible disturbance from my presence were used for quantitative analysis.

Selecting all *kewalbuga* males and married villagers of both sexes who lived in the Ginar main settlement as the subjects of the two types of activity survey presented two problems. The first was the omission of children. It was nearly impossible to pursue the children's frequently changing activities although, whenever they followed adult villagers, they were checked in the same manner as belongings. Neither the activity-tracing survey nor the time-recording survey disclosed any significant contribution to subsistence by the children (here, such activities as child care and cooking were excluded). Thus, the omission of the children's activities was not serious. This generalization as to children's lack of participation in subsistence activities seems to be supported by Minge-Klevana's (1980) recent review of time-allocation studies, where she, citing Nag, White and Peet's (1978) comparative study in Java and Nepal as well as a good deal of other literature, concluded that the work input by children in hunting-gathering societies is more limited than in peasant societies and therefore need not be included in measures of family labor. The Oriomo Papuans' basic life style is rather akin to hunting-gathering in terms of the above dichotomy. The second problem, the omission of males belonging to the *nanyuruga* age-grade, arose accidentally since, of the three *nanyuruga* men, two were Gimurugi-dwellers and one was a dual dweller of Ginar and Gimurugi. According to my limited observations, their activity pattern was more similar to *miid* than to *rugajog*, so the present analysis treated *miid* and *nanyuruga* as a single category in cases where the activities of *nanyuruga* men were analyzed.

If the three kinds of investigation mentioned above had been conducted concurrently, there would be greater validity in evaluating the interrelationships among different data. Since my fieldwork data were collected by a single person, i.e., myself, this, however, proved impossible. Accordingly, the time-recording survey and food consumption survey were set separately; the time allocation for these investigations is

Table 23. Survey periods for time-recording and food-consumption

month	season	survey
1971 Jul		
Aug		
Sep	dry	
Oct		] dry season's time-recording survey (Oct 11–23)
Nov	↓	] dry season's food-consumption survey (Oct 25–Nov 5)
Dec	↑	
1972 Jan		
Feb	wet	] west season's food-consumption survey (Jan 31–Feb 11)
Mar		] wet season's time-recording survey (Feb 28–Mar 11)

7. A boy shooting at a small lizard

8. Four sets of bows and arrows

9. A man who has just shot an arrow at a grass-wallaby (*Wallabia agilis*) running away

10. A man shooting an arrow of *dupa* at a bird perching on a tree branch

11. A man waiting for animals to run out of the burning area on a group hunt

12. Hunters at the final stage of a group hunt

13. The entrance to a sago grove in the swamp

14. A fallen sago palm, whose surface wood has been partially removed

15. Women pounding sago pith with *abol*

16.

17.

16. A woman washing out sago flour: the suspended sago starch running into a vessel with water

17. Powdery sediment of sago flour

18. Firing trees, shrubs and grass on the new garden plot

19. A woman planting taro shoots in the new garden

20. An old garden with perennials left growing

21. Men dissecting cassowary

22. Firing "casserole," the lower part consisting of logs and the upper part of small pieces of pisolitic laterite

23. "Casserole" covered with the bark of *ti* trees (*Melaleuca* sp.)

24. The cooked meat

shown in Table 23. The activity-tracing survey was made under a sporadic time allocation throughout the whole duration of my study period, mainly in September, October and November 1971 and February and March 1972.

Food Consumption

The data from the food consumption survey for twelve successive days in each season are shown separately by household and by season in Table 24, where Thomson's (1954) man-value coefficients for adjusting individual differences in age and sex (see Appendix 7) are applied. The wasted portion of consumed food was immediately measured or estimated by measuring several samples of the same species of food plants or animals on separate occasions. Table 25 shows the proportion of energy and protein intake for each foodstuff, based on mean food consumption for the 17 persons of the four households studied by using the analyzed nutrient values of our food samples, and to some extent the values from two food composition tables (Hodges, Fysh and Rienits 1947; McCance and Widdowson 1960), as shown in Table 20. As these tables (Tables 24 and 25) demonstrate, production of sago, slash-and-burn horticulture and hunting are the major subsistence activities and provide the bulk of the diet. Of the food items, sago is very prominent in terms of quantity of consumption and as an energy source since plant foods rich in carbohydrate and thus forming a staple food group are evaluated in terms of their energy supply. A nutritional evaluation based on energy intake cannot be applied to animal foods, simply because their dietetic significance derives from protein supplied. In summary, about 60% of the villagers' energy comes from sago flour and about 60% of their protein from game animals.

Table 26 shows the average intake of energy, protein and fat from all foodstuffs. The energy intake was high, taking into account the body weight of the villagers (the mean weight of 25 adult males was 57.8 kg with a standard deviation of 6.0). On the other hand, the intake of protein and fat was rather small. This suggests a unique dietary pattern in which the people must consume a large amount of carbohydrate to acquire the necessary protein and fat. This unique pattern is mostly attributable to the unbalanced composition of the

Table 24. Average daily per-capita food consumption[a] of the four households[b] (g)

	Household No.	man-value coefficient	sago	coconut	taro	banana	papaya	pine-apple	other crops	wild fruits	tree leaves	meat	fish	tree ants
dry season	1	2.95	485	58	756	495	217	139	27	149	c	98	0	0
	6–1	2.85	811	0	267	274	253	0	0	284	0	140	0	c
	6–2	2.30	1,174	61	183	0	239	0	0	230	0	100	0	c
	14	2.60	908	104	188	262	227	223	38	192	0	142	0	c
	19	0.85	941	82	141	271	400	94	35	141	c	106	0	c
	average	11.55	831	56	348	273	246	93	18	208	c	119	0	c
wet season	1		231	41	0	912	203	20	0	27	c	142	27	0
	6–1		1,235	14	0	379	246	0	0	14	c	81	0	0
	6–2		1,270	26	0	652	204	0	22	35	0	135	0	0
	14		1,138	54	0	642	181	42	0	4	0	242	0	0
	19		694	82	0	1,518	388	94	0	71	c	129	59	0
	average		923	37	0	713	223	22	4	23	c	147	11	0
average			878	47	174	493	134	57	11	116	c	133	6	c

[a] The amount of consumption was calculated as that for a single adult male person according to Thomson's (1954) man-value coefficient.
[b] Household No. 6 is divided into two meal units because of the two polygynous wives.
[c] Negligible amounts.

Table 25. Proportion of energy and protein intake by all foodstuffs

Energy

	dry season	wet season	average
sago	53.8	63.5	58.5
coconut	1.7	1.2	1.5
garden crops	27.8	29.9	28.9
(taro)	(13.3)	(—)	(6.9)
(banana)	(9.5)	(26.4)	(17.7)
(papaya)	(3.3)	(3.1)	(3.2)
(pineapple)	(1.3)	(0.3)	(0.8)
(others)	(0.4)	(0.1)	(0.3)
wild fruits	13.3	0.8	7.2
meat	3.3	4.3	3.8
fish	—	0.3	0.1

Protein

	dry season	wet season	average
sago	7.3	7.9	7.6
coconut	1.3	0.8	1.1
garden crops	24.9	19.2	21.9
(taro)	(14.0)	(—)	(6.8)
(banana)	(6.7)	(16.6)	(11.7)
(papaya)	(2.7)	(2.3)	(2.5)
(pineapple)	(1.1)	(0.2)	(0.7)
(others)	(0.4)	(0.1)	(0.2)
wild fruits	8.9	1.5	5.1
meat	57.5	67.1	62.4
fish	—	3.6	1.8

nutrients in sago, 100 g of which contains 0.4 g protein, 54.8 g carbohydrate, 44.5 g water, and 220 kcal energy; according to the literature for food composition (Hodges, Fysh and Rienits 1947; Peters 1957; McCance and Widdowson 1960), the corresponding figures are 0.1–0.2 g protein, 71–83 g carbohydrate, 17–27 g water and 285–349 kcal energy, showing higher values for carbohydrate and energy, and a lower value for protein. While it is the case that staple plant foods, tubers or cereals, are generally regarded as energy sources, sago is distinguished as the most common source. In dry weight ratio, the protein content of sago is one-fifth to one-tenth that of the bananas or tuberous crops analyzed in this study. Thus this staple must be supplemented by other foods, especially animal food rich in protein. This view has been

Table 26. Average daily per-capita[a] intake of major nutrients and energy

	dry season	wet season
energy (kcal)	3,399	3,208
protein (g)	44.9	47.1
fat (g)	38.7	12.8

[a] Adjusted to the value consumed by a single adult male in the same way as in Table 24.

emphasized by various researchers (e.g., Oomen 1971; P. K. Townsend 1974). Many game animals acquired during the hunts provide this supplement in Wonie. In this sense, the villagers' traditional opinion that sago starch combined with game animals forms a proper diet is reasonable.

Here, the amount of animal food consumption will be assessed again. As mentioned above, in general terms the villagers' total protein intake is at the marginal level, although an average of 133 g of consumed weight of animal per day per head is comparable to the corresponding 147 g among the Kalahari San (formerly called Bushmen), who are one of the few surviving hunting-gathering populations (Tanaka 1976). Thus, it may be concluded that the hunting of the Oriomo Papuans provides animal meat to an extent which is common among "proper" hunter-gatherers in the tropics.

As shown in Table 24, there was a fairly considerable difference in foodstuffs consumed between households although, in my view, this difference might be largely due to the short period of the investigation. It is noticed that variations in consumed amounts of sago and garden crops produced the most easily distinguished inter-household differences. The menu of individual households changes in accordance with the progress of sago-making and the harvesting of various garden crops, both of which differ from household to household. Every household obtains sago starch every ten days or two weeks, and therefore the proportion of sago consumed is the highest just after sago-making and tends to lessen until the next sago-making, with a counter-balancing increase in garden crops. Meat consumption also tends to fluctuate from day to day according to when animals are killed and, thereafter, whether the animals are shared with the household in question. Judging from average daily consumption, however, inter-household variation was much smaller in meat consumption than in plant food consumption. This contrast between plant and animal foods is a product of their harvesting and food-sharing systems. The plants are, as a rule, harvested by each household and are seldom shared with others, while animals tend to be impartially distributed among all households.

In terms of seasonal differences in average food consumption, of the major foods, sago, coconut, papaya and animal meat were available to an identical extent in both seasons; taro was available exclusively in the dry season; and bananas became increasingly available in the wet season. The harvesting season of taro and other tuberous crops depends upon botanical factors, although strictly speaking the physiology and ecology of Oriomo taro and other tubers in terms of seasonality still remain relatively unknown. There is a high probability that bananas are harvested to a similar extent in both seasons. I failed to determine whether variety (local kind)-specific harvesting seasons exist, but if any kind of banana fruit can be cropped from transplanted shoots after maturation periods of similar length, it is reasonable to assume that the people harvest more bananas in the wet season because taro is unavailable then. In spite of these seasonal variations, each of the three food groups, i.e., sago, garden crops and meat, which comprise the bulk of the food supply in this community is almost equally available in both seasons. The constant year-round intake of both animal and plant foods, which is of nutritional-physiological significance, is most valuable.

Problems concerning the stability of the food supply can be summarized by four points. First, animal foods are consumed fairly stably in terms of season, and every household studied consumed a similar amount of animal meat during the two twelve-day periods. Second, the year-round availability of sago starch was reflected in its similar consumption during wet and dry seasons, although the inter-household day-to-day variation was great. This leads to the third point. As mentioned previously, the chance shortage of sago was compensated for by other staples from horticultural gardens. The garden crops not only enrich the variety of the menu but also contribute to stabilizing the food consumption. The final point is concerned with three minor food-getting activities, i.e., coconut exploitation, collecting and fishing, which have not yet been discussed. Coconuts and fish occupy a very small part of the total consumption. Of the wild plants, the *Canarium* fruits were thought to be consumed to a larger extent than usual during the dry season study because four of the fewer than ten (total number during the whole 1971 season) *Canarium* trees were felled during the twelve-day food consumption survey period. This suggests that the wild plant foods provide a small amount of food in the long-run and are subject to seasonal factors which cannot be controlled by the people. Thus, we can conclude that these minor food-getting activities do contribute to the enrichment of their menu but contribute slightly to the stability of their food consumption.

Time Allocation

The length of time spent in a single activity differs considerably from one kind of food-getting activity to another. Figure 16 shows the percent distribution of time by two-hour intervals for daytime activities and by the number of days for overnight activities. The number of activities were counted in terms of person-time; for example, when a single activity was carried out by two people it was counted as two person-times. The time of a single activity includes walking to and from the village and working places. This figure shows that horticulture and individual hunting activities are fairly concentrated in shorter work periods. About 50 percent of all single activities

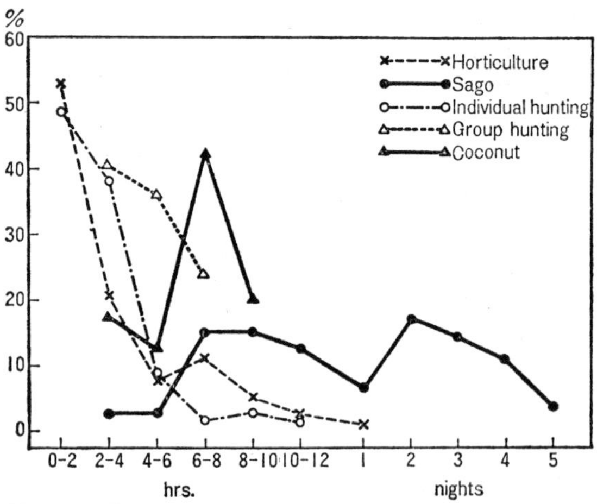

Fig. 16. Distribution of Labor Time for Single Activities.

involved in both these categories fell into the 0–2-hr category, and more than 70 percent of horticulture and more than 85 percent of individual hunting activities fell between the 0–2- and 2–4-hr categories.

The distribution pattern for individual hunting reflected two types of short-time activity. The first was the search for grass-wallabies and bush-wallabies during feeding hours, which necessarily takes place for a short period of time in the early morning or evening. The second type of hunting practice took place around the village and was done when the hunter had nothing else in particular to do rather than when he earnestly wanted the game. Consequently the yields from this type of hunting were much smaller. Individual hunting activities requiring long periods of time usually linked with travels to remote places, and might at the same time include inspection of the sago groves. In contrast to individual hunting, group hunting required from 2–4 to 6–8 hrs, simply because this method involves burning grasses. Excluding walking between the village and the hunting ground, and preparation and treatment of the game killed, it takes at least one hour and a half to accomplish this type of hunting.

The short-time categories, especially 0–2 hrs, spent in horticultural activity usually involved obtaining small amounts of crops. The harvest of garden crops is normally accomplished in small stages. Horticultural activities requiring a long time include cutting trees and shrubs for new gardens, making fences around the garden plots (though fences were not so frequently constructed) and weeding new gardens. For such horticultural labor, a husband and his wife or wives usually go together, frequently accompanied by their children. While the garden construction is accomplished with the direct cooperation of a husband in cutting big trees and his wife or wives in cutting shrubs and grass, the other two kinds of labor do not require cooperation. During men's fence-making time, the women are, in many cases, engaged in planting cuttings or shoots. During women's weeding time, the men usually inspect the surroundings for unspecified purposes and/or rest.

The variation in the length of time spent on a single activity is associated with how many people are involved. In order to analyze both time and people involved, the former was classified into three categories depending on length, and the latter

Table 27. Distribution of single activity time for horticulture by membership (number of observations in terms of person-time and percent)

hours	individual			married couple	two or more women
	man	woman	total		
0–2	59 (64.1)	124 (75.2)	183 (71.2)	44 (27.5)	8 (28.6)
2–6	24 (26.1)	38 (23.0)	62 (24.1)	54 (33.8)	14 (50.0)
6–	9 (9.8)	3 (1.8)	12 (4.7)	62 (38.8)	6 (21.4)
total	92	165	257	160	28

into three categories since observed participation was of three types: (1) an individual, (2) a married couple, or a husband and his wife in a monogamous family, or a husband and one or two of his wives in a polygynous family (all Wonie polygynous families involved two wives), although in a very few cases, two couples, and (3) two or more women who went together to gardens located close together (Table 27). This analysis shows that the longer the work took the lower the proportion of activities performed by an individual and the higher the proportion of participation by married couples. Three other observations can be added to this analysis. First, a male group is not organized for horticultural labor, even though plural married couples sometimes go separately to their partitioned garden. Second, the proportion of an individual's long-term labor is higher for men, reflecting the lengthy labors of clearing big trees and fence-making. This contrasts with the third observation, that is, that groups of women unaccompanied by their spouses are frequently not organized, but this type of group organization assures, even partly, the women's lengthy labors when their spouses are engaged in other activities like hunting.

The distribution pattern for time spent in sago exploitation, i.e., sago-making, shows a tendency opposite to the patterns in horticulture and individual hunting. Most sago labor fell into the 6–8 hr category or longer (Fig. 16). A more prominent feature is that more than half (exactly 51.8%) of sago labor required overnight stays in the sago groves. The people prefer to stay in the working place when it is more than an hour's walk from the village. However, the proportion of overnight activities was low in the wet season because the people hesitate to stay in poorly-structured huts and also because they tend to exploit sago groves relatively close to the village. On overnight stays, each married couple is usually accompanied by their children and bring daily necessities, which are in fact only cooking utensils. In the working sites, women are almost totally involved in sago-making, while men not only perform their duties, i.e., cutting down the palm trees and removing the surface wood, but also inspect the sago stands in the area and/or go hunting. The children spend their time in the same way they do in the village.

The activities involved in coconut utilization show a narrow distribution, with a peak of 6–8 hrs. This pattern is recognized as an extension of the distribution pattern in sago-making from which the overnight activities are excluded. This feature reflects the similarities and dissimilarities between sago-making and coconut-fetching. Coconut-fetching, like sago-making, is performed by a group and generally in places remote from the village, but does not require as much time as sago-making does. Fishing and collecting activities are not shown in the figure because of the paucity of my observations. However, the distribution pattern of fishing time is estimated to have been similar to that of group hunting, and collecting time may always have taken 0–2 hrs.

The daily rhythm of food-getting activities, here represented by the proportion of active persons to the total number of persons available, irrespective of sex and season, is shown in Figure 17. Sexual differences and seasonal variations for food-getting activities in the daily time course were negligible.

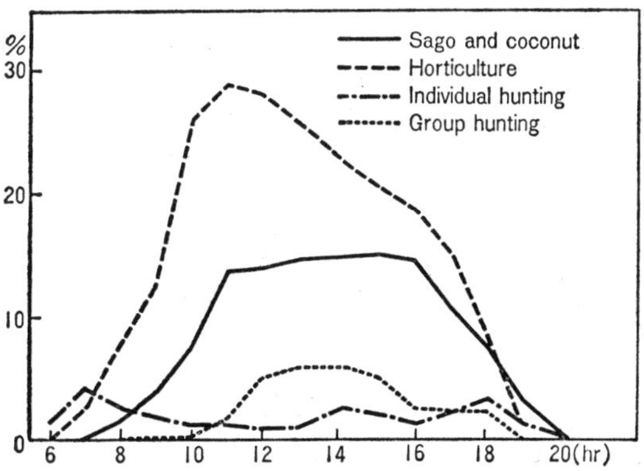

Fig. 17. Daily Activity Rhythm.

Active persons refer to those who were out of the village during each activity, although strictly speaking it was not certain whether the people at the work site were actually working. In this figure, sago utilization and coconut utilization are indicated *en bloc* because of their identical pattern, and fishing and collecting are omitted because few activities were involved. If we standardized the number of available persons as the denominator in the calculation of the proportion, the frequency of individual hunting each hour might roughly double, since only males are engaged in this activity, and that of group hunting might quadruple since this activity is restricted to the dry season.

The rhythm of group hunting contrasts with that of individual hunting. The former has a peak from 10: 00 to 15: 00, the hours favorable for burning grasses. Individual hunting has two peaks, early morning and evening, because, according to the villagers, game animals, especially grass-wallabies and bush-wallabies, usually feed on grasses and tree leaves at these times, and thus it is easier to find the animals. Whatever the reason, the daily rhythm of hunting activity coincides with the ecology or habits of the animals and/or the daily dynamics of environmental conditions. In contrast, the rhythm of sago utilization, coconut utilization, horticulture, and perhaps collecting, i.e., activities involving plant foods, seem to be determined only by factors relating to the people themselves. In spite of the fact that various plant-getting activities in the Oriomo subsistence complex involve domesticated, semi-wild and wild species, this does not affect the people's daily rhythm.

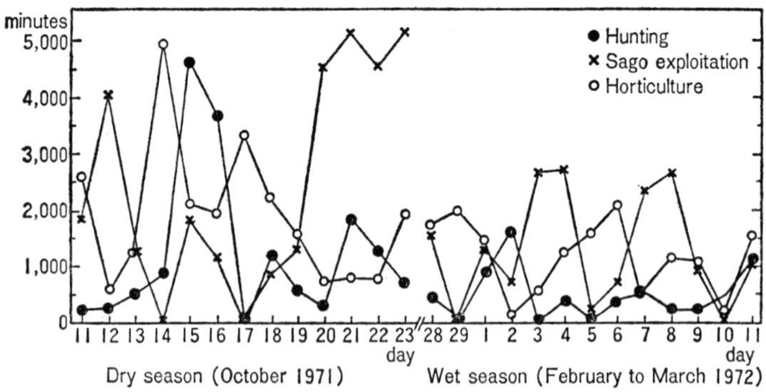

Fig. 18. All Villagers' Time Spent Each Day in Major Food-Getting Activities.

Thus, the major complex of activities, consisting of hunting and various plant procurements, condition the people's daily time allocation in the same way that hunting and gathering condition the time allocation of hunter-gatherers.

This point of discussion will be assessed again using the data of day-to-day fluctuations of time spent in certain activities. Figure 18 shows total time use by all adults for the major activities of each day during the 26-day survey periods for the dry and wet seasons. Time calculated for this figure includes walking back and forth between the village and the working sites. In cases which involved overnight activity, generally during sago-making, an estimation was made to calculate the time. According to my observations in the activity-tracing surveys of such cases, the average proportion of time spent at the working site to the total time between 6:00 and 20:00 was 60 percent, so this proportion was applied to the estimation of actual working time during out-of-village hours.

Each activity time fluctuated from day to day, and the statistical calculations indicated that the coefficients of variation (standard deviation/mean × 100) among the 26-day records were 131.4, 88.2 and 68.1 for hunting, sago exploitation and horticulture, respectively. The time correlations for the three major activities in each day showed a clear trend (Table 28). Time for sago exploitation correlated negatively with horticultural time at a significant level (T-test, $p<0.01$), whereas hunting time had no significant correlation with the

Table 28. Simple and partial correlations between times spent in subsistence activities

$r_{XY}=-0.399$*	$r_{XZ}=0.052$	$r_{YZ}=0.068$
$r_{XY \cdot Z'}=-0.399$*	$r_{XZ \cdot Y'}=0.052$	$r_{YZ \cdot X'}=0.077$

here, X: time for sago exploitation,
Y: time for horticulture,
Z: time for hunting,
X′: total time for sago exploitation, coconut exploitation, fishing and collecting,
Y′: total time for horticulture, coconut exploitation, fishing and collecting,
Z′: total time for hunting, coconut exploitation, fishing and collecting.

* Significant difference at a probability of 0.01 (T-test statistics).

Table 29. Per-day per-person time use by all adults (minutes; percent in parentheses) and the dry season-versus-wet season ratio

	hunting	fishing	horti-culture	sago exploi-tation	coconut exploi-tation	collect-ing	total
dry season	36.8 (11.6)	0.4 (0.1)	114.4 (36.0)	146.3 (46.1)	18.7 (5.9)	1.0 (0.3)	317.6
wet season	14.1 (7.5)	6.2 (3.3)	69.7 (37.0)	77.3 (41.0)	21.1 (11.2)	0	188.4
average	25.5 (10.1)	3.3 (1.3)	92.1 (36.4)	111.8 (44.2)	19.9 (7.9)	0.5 (0.2)	253.0
dry-wet ratio	2.60	—	1.64	1.89	0.89	—	1.69

other two. Even if the idea of partial correlation was applied, the interrelations of the three showed similar levels. This interrelationship is relevant to their subsistence system as a whole. There are two common factors in horticulture and sago exploitation. One is the exploitation of plant food resources, and the other involves the married couple as the most basic unit of labor. These are in contrast to hunting which is practiced only by men. Thus, synthesizing the interrelationship of time use with the nature of the activities, it is reasonable to conclude that the major activities revolve around plant-getting and hunting, and that these two types of activity are fairly separately organized. In general terms, this dichotomous classification demonstrates one of the essential differences between the procurement of animal resources and that of plant resources.

Table 29 shows per-day per-person time spent in each food-getting activity by age-grade, according to season; here, working time in overnight activities was determined as mentioned in the analysis of Figure 18. The data demonstrate that three major activities, sago-making, horticulture and hunting, occupy on the average more than 90 percent of the total food-getting activity time. Per-day per-person time used for all food-getting activities by all adult persons, i.e., married men and women and *kewalbuga* males, averages 253 minutes, or about four hours and a quarter. For a comparative assessment of this value, corresponding times (working time plus walking time between settlement and working sites) in other populations need to be reviewed. Among the San (≠Kade Bushmen) of the Kalahari, per-day per-person activity time averages

279 minutes (Tanaka 1980:77), and among the Australians inhabiting the Arnhem Land, it is four hours or less (including time for food preparation) (McCarthy and McArthur 1960).

In spite of a greater variation in food-getting time in populations which are generally involved in horticulture, some show a similar level. Among the Bemba of northern Rhodesia, who depend on hoe cultivation, fishing and hunting, males use 3.4 hours and females 6.0 hours (Richards 1939:390); among the Machiguenga of the upper Amazon, who depend on slash-and-burn horticulture, gathering and hunting, males use 4.4 hours and females 1.8 hours (Johnson 1975); and the Hanunóo of the Philippines, who are basically rice cultivators, spend 230 minutes on the average for both sexes (Conklin 1957, with my additional calculations). Some other horticultural societies show longer activity times, although they appear to be largely subject to a cash economy. The similarity of the time spent in food procurement (about four hours) is one of the essential characteristics of people who rely on local products through hunting, gathering or horticulture for their daily livelihood. This may parallel Sahlins's (1968, 1972:41–69) hypothetical theory of the pristine "affluence" in the hunting-gathering and slash-and-burn horticultural economies, i.e., underuse of resources, underuse of labor-power and underuse of time for actual labor. This characteristic is a source of stability in such peoples' adaptation, including the Oriomo Papuans'.

We turn now to the next problem. The time used for food-getting activities differs to a considerable degree by season. The dry-season/wet-season ratio in total time reaches 1.69, and the time ratio for each of the three major activities, i.e., sago utilization, horticulture and hunting, is 1.89, 1.64 and 2.60, respectively. There are several possible reasons for this difference. The differences in horticulture and hunting are largely due to environmental (or climatological) conditions. Both group hunting and garden-making, which is the most time-consuming component in horticultural activities, can be practiced only in the dry season when the burning of grasses and/or trees is possible. In other words, plant ecology conditions hunting and horticultural activities as a whole. A further minor reason for increased hunting in the dry season comes from the fact that *kewalbuga* males, whose hunting practice is more of the nature of training or leisure, prefer individual hunting in the dry season to that in the wet season because the hunting trips in the wet season are more uncomfortable. Sago-making is directly affected by climatic conditions. Frequent squalls in the wet season tend to make sago starch production impossible since the starch may run off in the process of pounding and washing out. Also, overnight stays at the sago grove tend to involve uncomfortable conditions in the wet season. To cope with this, the people work in nearby sago groves more frequently, and inspect their groves more earnestly to avoid cutting palm trees filled with little starch. In fact, they tend to work in sago groves where several mature stands exist together.

In spite of the seasonal differences in time use for each of the major activities, the yields do not differ greatly; for example, sago starch must be produced in greater quantity in the wet season when tuberous crops are seldom available (cf. Table 24). In other words, these seasonal differences lead to ecological conditions which affect the people's survival; that is, their subsistence adaptation is more severely conditioned or more skillfully maintained in the wet season than in the dry season. This viewpoint is of great validity in considering human ecological systems; for example, the carrying capacity of the land, or the balance between the human population and the land area or the resources in it, should probably be assessed in the wet season, although my data could not fully demonstrate this.

The final theme concerning time allocation is individual variation. As shown in Table 30, great differences in activity time based on age and sex provide another focus of interest. The total time use per day ranged from 110 minutes for *kewalbuga* to 306 minutes for *rugajog*. *Kongajog* worked at 90 percent of *rugajog* time, and *miid* and *nanyukonga* at 70 percent of it. It might be reasonable to suppose that males spent more time because of their exclusive participation in hunting. However, it was also true that the women spent much more time on such domestic affairs as cooking and child care. I suppose that on the average the men's hunting time approximates or falls behind that of the women's domestic work. In contrast to the smaller variation among all married age-grades of either sex, the comparative rate of unmarried

Table 30. Average per-day per-person time spent in food-getting activities (minutes). Thirteen-day records for each season were averaged, and the number of persons of each age-grade is shown in parentheses following the age-grade's name.

A) Dry season

	group hunting	individual hunting	fishing	horticulture	sago exploitation	coconut exploitation	collecting	total
miid (2)	20	0	0	171	127	0	0	318
rugajog (12)	48	23	0	133	169	19	0	392
kewalbuga (9)	77	19	15	10	16	0	9	137
nanyukonga (4)	—	—	0	104	117	15	6	242
kongajog (14)	—	—	0	125	187	23	0	335

B) Wet season

	group hunting	individual hunting	fishing	horticulture	sago exploitation	coconut exploitation	collecting	total
miid	—	0	4	130	92	0	0	226
rugajog	—	43	0	75	74	24	0	216
kewalbuga	—	19	3	9	40	0	0	71
nanyukonga	—	—	9	60	89	24	0	182
kongajog	—	—	12	78	88	32	0	210

C) Average (% in parentheses)

	hunting	fishing	horticulture	sago exploitation	coconut exploitation	collecting	total
miid	10 (3.7)	2 (0.7)	151 (55.5)	109 (40.1)	0	0	272
rugajog	57 (18.6)	0	105 (34.3)	122 (39.9)	22 (7.2)	0	306
kewalbuga	58 (52.7)	9 (8.2)	10 (9.1)	28 (25.5)	0	5 (4.5)	110
nanyukonga	—	5 (2.3)	82 (38.5)	104 (48.8)	19 (8.9)	3 (1.4)	213
kongajog	—	6 (2.2)	102 (37.2)	138 (50.4)	28 (10.2)	0	274

kewalbuga drops to 35 percent. This striking feature explains the very short time they spent on sago utilization and horticulture: more than 50 percent of the *kewalbuga*'s food-getting time was spent hunting, although their hunting time was identical to that of *rugajog*. Furthermore, their hunting tended towards group hunting which requires less skill and work loads compared to individual hunting. The fact that the *kewalbuga* males seldom work in sago utilization and horticulture, which provide the bulk of the villagers' diet, means that they depend heavily on the married members for daily sustenance, or are not seen by the married members as making an important contribution to the villages' current lives.

In terms of comparisons among males and those among females, two different characteristics are observed. Prominently featured among the males are not only *kewalbuga* but also *miid*, who tend to retire from hunting, sago-working (actually cutting sago trees), and coconut exploitation (actually climbing coconut trees), which require heavy work loads. As compensation, the *miid* spent much time in horticulture. In fact, these old men performed the women's tasks, planting and weeding in the gardens, more frequently than *rugajog* men. Another compensation is seen in the *miid*'s fishing, something which was never done by the *rugajog*. The fishing time registered in Table 30 was spent by one of the *miid* (Indiv. No. 1-1) who had given up his bow and arrow. On the contrary, the differences between the female *nanyukonga* and *kongajog* are very small. Thus, the sex and age differences for time spent in food-getting activities can be summarized as follows: with increasing age, women's work changes only slightly in quantity while men's changes not only in quantity but also in quality.

Labor Efficiency

The time spent in subsistence activities should be assessed not only by absolute length but also by relative productivity. Labor productivity, or the balance between labor input and output for each activity is, of course, an essential human ecological condition. In my view, however, the labor productivity is of particular importance in two different aspects, i.e., inter-activity comparison and inter-individual comparison.

Two barriers hamper identification of inter-activity labor productivity for the data of the present study. The first is the difficulty of measuring output, that is, the return of labor input, even if the input is indicated by labor time, not by energy consumption, which is more valid in a nutritional-physiological sense. Carneiro (1957) and his followers calculate the amount of energy (or calories) involved in the products per unit of labor as this indicator. This measure possesses considerable validity in comparison between plant-food-getting activities, since plant foods, especially staples such as tubers and cereals, contain a large amount of carbohydrate and act mainly as energy sources. Thus, the inter-activity comparison of productivity may have to be used to treat plant-getting activities exclusively; however, the productivity of hunting will also be treated in this way here for a crude comparison.

The second barrier derives from the nature of labor. In the Oriomo subsistence system, procurement of wild foods as well as sago and coconut requires little or no labor input other than harvesting, while horticulture involves making a new garden, planting, weeding and harvesting, which continue in succession. In addition, horticultural labor requires caring for various kinds of crops at the same time. For surmounting the second barrier, two types of comparison have been undertaken using energy return per labor time as the indicator of productivity. The first is a comparison among the productivity of sago exploitation, *Canarium* fruit collecting and hunting. The input-output record for the former two kinds of activity is based on my observations in the activity-tracing surveys. Unworked time in the working places, for example, for resting and cooking, is excluded from the calculations. On the other hand, data for hunting consist of the total time spent in hunting by all adult males recorded in the time-recording survey, and the total game animals acquired in the same survey period because during the activity-tracing survey my presence on hunting trips disturbed the hunters and might have frequently allowed animals to escape, and consequently the records failed to yield sufficient labor efficiency. On the other hand, according to my observations of various hunting trips, the hunting activities went on without significant breaks. Therefore, the hunting time observed in the time-recording survey can be considered the net activity time.

The second comparison is made for identifying horticultural efficiency. For this purpose, horticulture and sago-making, both of which were frequently practiced, are compared in the time spent in all stages of activity as shown by the time-recording survey (Table 29 or 30) and the quantity of energy intake as shown by the food consumption survey (Table 24).

Table 31. Average input-output of horticulture, sago-making, collecting of *Canarium vitiense* fruit, and hunting

	horticulture	sago	*Canarium*	hunting
A-type calculation				
product (kg)/labor (hr)	—	1.9	2.9	1.3
energy (kcal)/labor (hr)	—	4,290	6,550	1,250
B-type calculation				
energy (kcal)/labor (hr)	980	2,000	—	—

A-type calculation is made with direct observation data for sago-making and *Canarium* collecting, and with time-recording surveys for hunting; B-type is based on the data from time-recording and food-consumption surveys (see text).

Taking sago exploitation as a link with horticulture on the one hand and with *Canarium* collecting (and hunting) on the other, a synthetic comparison is possible, even if based on crude indications.

Table 31 shows the results of the two types of comparative calculation. Comparing the results of eight time observations of sago-making with those of five time observations of *Canarium* collecting (in the upper part of Table 31), the average energy return per man-hour of labor for the former is around two-thirds that of the latter, i.e., 4,290 kcal versus 6,550 kcal. The lower productivity for sago-making appears to derive from the complexity of the technique, which includes pounding and washing out. Even if the labor efficiencies are recognized as identical, it can be at least concluded that the "elaborateness" of technology involved in sago-making does not make for higher productivity, in comparison with *Canarium* collecting, which consists of felling a tree and plucking off the fruit. Calculating the weight of animals killed per time spent in hunting during the 26-day time-recording survey period, we find that the efficiency of energy is rather low in comparison with sago-making and *Canarium* collecting although, as stated above, the comparison of energy returns is less meaningful.

In the second comparison, shown in the lower part of Table 31, the activity time used as the denominator for measuring labor productivity includes all time spent out of the village, a considerable portion of which was spent travelling between the settlement and various working places and in resting or talking at the working places. Men, in particular, used a smaller portion of their time for actual work when at the gardens or sago groves. Similarly, if not to the same extent, the length of time women spent in the gardens was only a crude indication of the time spent in horticulture. Bearing these reservations in mind, the data indicate that one man-hour of labor provides about 1,000 kcal in horticulture and 2,000 kcal in the exploitation of sago. In general terms, it can be concluded that the exploitation of sago is more productive than horticulture.

Combining the above two comparisons, the three plant-food-getting activities are placed in decreasing order of labor productivity as follows: *Canarium* collecting, sago exploitation, and then horticulture. In my view, this order parallels the increasing technological complexity and necessity of preparation before direct harvesting. The idea that the labor efficiency of gathering is not necessarily behind that of cultivation may be widely accepted by anthropologists who consider the early stages of agriculture in an evolutionary perspective (e.g., Bronson 1977). Even if we ignore *Canarium* collecting, which provides a small amount of food for the villagers in terms of yearly output, the superiority of sago exploitation to horticulture is recognized as one of the important ecological factors in the subsistence system of this population. Also emphasis should be placed upon the coexistence of various kinds of activities, whose productivities differ greatly. While labor efficiency occupies a basic place in the labor settings of most societies, especially those functioning under a cash economy, the Oriomo's system depends largely upon how food resources can be exploited by their technology. This characteristic should be included as a component in the pristine subsistence system.

The inter-individual labor efficiency of the major activities, sago-making, collecting and horticulture, all of which commonly procure plant foods, showed small individual differences and slight day-to-day fluctuations when time was used for the same purpose. These characteristics of plant-food-getting activities, most of which are performed by women, are very conducive to a stable supply of the indispensable portions of the people's daily food. As I pointed out elsewhere (Ohtsuka 1977a; also Watanabe 1971), hunting and fishing, which seek animals *ferae naturae*, depend subtly on biotic and abiotic con-

Table 32. Hunting time used and animals killed by all villagers on each day of two 13-day periods

Dry season

date	individual hunting			group hunting		
	no. of hunters	total time (minute)	animals killed (no.)	no. of hunters	total time (minute)	animals killed (no.)
Oct 11	2	190	grass-wallaby (2)	0		
12	3	230	pigeon (2)	0		
13	5	500	grass-wallaby (2) pigeon (1)	0		
14	4	830	bandicoot (1) small bird (1)	0		
15	2	90		13	4,550	pig (2) grass-wallaby (1) bush-wallaby (5) bandicoot (6) small lizard (2)
16	3	370	cassowary (1)	11	3,290	grass-wallaby (3) bush-wallaby (3) bandicoot (2) monitor (1)
17	0			0		
18	0			8	1,210	bush-wallaby (3) bandicoot (4)
19	4	530	small bird (2)	0		
20	1	240				
21	1	360		4	1,480	scrub-wallaby (1)
22	2	210	grass-wallaby (1) pigeon (1)	5	1,030	
23	2	660	cassowary (1)	0		

Wet season (individual hunting only)

date	no. of hunters	total time (minute)	animals killed (no.)	date	no. of hunters	total time (minute)	animals killed (no.)
Feb 28	3	380		Mar 6	6	370	bush-wallaby (2) pigeon (1)
29	0						
Mar 1	5	860		7	2	490	
2	13	1,660	bush-wallaby (7)	8	2	190	
3	0			9	1	180	
4	3	380	bush-wallaby (1)	10	3	450	pig (1)
5	0			11	4	1,100	grass-wallaby (1) bush-wallaby (1)

ditions in the local environment and therefore tend to bring about a great day-to-day instability and individual variation in the input-output balance or in labor efficiency. From my observations, this view applies to the hunting activities of this population, although fishing activities were performed too seldom to be assessed from this aspect. Table 32 shows the case record for each day's hunting time and its yields during the time-recording survey period. Along with a great day-to-day fluctuation in time spent hunting by all hunters, these data strongly suggest the day-to-day instability of hunting efficiency.

It is doubtless that the labor efficiency varies to a great extent from hunter to hunter. Practically speaking, however, individual hunting efficiency is compared here by age-grade due to the paucity of data. The available data were gathered on 39 days consisting not only of two 13-day periods for the time-recording survey in October 1971 and February to March 1972, but also of another 13 days in the semi-dry season between 19 and 31 July 1971. The average and total values of input-output throughout the observation periods are shown in Table 33, in which the hunters are divided into three groups by age-grade. A group of *miid* and *nanyuruga* is clearly distinguished from the other two, since their proportion of hunting time per observation time was very low and they actually killed no animals. In fact, throughout my whole study

Table 33. Hunting time and animals obtained by the three age groups

	miid and *nanyuruga*[a]	*rugajog*	*kewalbuga*
hunting time/observation time	0.005	0.084	0.070
animals (kg)/hunting time (hr)	0	1.06	0.45
no. of animals			
pig	—	3	—
cassowary	—	1	1
grass-wallaby	—	7	1
small wallabies[b]	—	19	2
bandicoots	—	9	2
monitors	—	—	1
small birds	—	—	3
lizards	—	—	1

[a] The Gimurugi-dwelling *nanyuruga* men were included when they went hunting from and returned to the Ginar main settlement.

[b] Including bush-wallaby and scrub-wallaby.

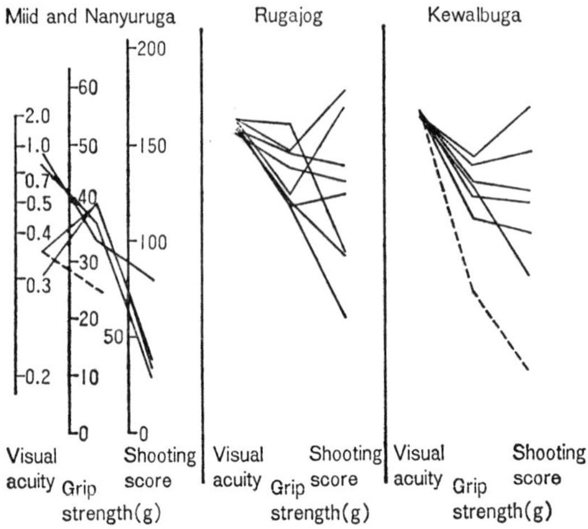

Fig. 19. Physiological Capacity and Scores in Archery-like Shooting Test.

period the *nanyuruga* men killed few animals and no *miid* man killed anything. The comparison between *rugajog* and *kewalbuga* shows a similar level of time engaged in hunting, but the labor efficiency of the former is more than twice that of the latter. Furthermore, the kinds of animals killed differ. Generally speaking, the proportion of animals which are killed with difficulty and whose meat the people like is higher on the list of products for the *rugajog* group, while the *kewalbuga* group's list consists of a great variety including trivial animals like small birds and lizards.

These differences in hunting productivity among the age-grades appear to be explained by their physiological capabilities and their past experience in hunting. Figure 19 shows individual hunters' visual acuity and grip strength (these data were mentioned previously and shown in Fig. 7 in a different form of illustration), and the scores on the shooting test which was held in the middle of January 1972; this shooting test followed the rules of an American archery round, and this method was established by Pope (1918) who studied the shooting capacity of Ishi, a famous American Indian survivor (cf. Kroeber 1964). The American round involves shooting thirty arrows at forty-, fifty- and sixty-yard distances. The target has a central ring of nine and half inches in diameter, and each of the outer four concentric rings is one-half this width. The values of hits are 9, 7, 5, 3 and 1 point, respectively.

Figure 19 suggests several matters involved in hunting skills. While considerable individual variations exist, it is clear that the visual acuity and grip strength of the *miid* and *nanyuruga* group lag behind the other two younger groups and that their shooting scores also show a lower level. It is reasonable to assume that the weakened physiological ability of the older people accounts for their low shooting score, their scanty hunting practice and their low labor productivity. One *miid* man (Indiv. No. 1-1), who had completely retired from hunting and is shown by a dotted line in the figure, is especially weak in grip strength. On the other hand, the youngest *kewalbuga* male (Indiv. No. K-9), who is also shown by a dotted line in the figure, demonstrates weak grip strength and a low shooting score. He entered the *kewalbuga* age-grade during my research period, September 1971, and his age at that time might have been 16 or 17 years. This suggests that young boys belonging to *yambuga* age-grade are not qualified hunters because of their less-developed physiques and inferior muscle strength (see also Figs. 6 and 7).

The similar levels of physiological ability and shooting scores for both *rugajog* and *kewalbuga*, and the doubled hunting productivity of the former to the latter, as shown in Table 33, are of great interest. This difference in hunting productivity is considered to be a product of the gradual advancement of hunting skill through practice. Thus, age factors relevant to hunting capacity are summarized as follows: young boys cannot be qualified hunters because of physical factors; the *kewalbuga* tend to attain full maturity in a physiological (or biological) sense but their hunting skills do not reach the *rugajog* level due to a lack of experience; the *rugajog* are the best hunters; and as the men grow from *nanyuruga* to *miid*, their physiological abilities become weak and consequently their hunting efficiency is lowered.

Labor Units

Of the main food-getting activities, only individual hunting and collecting *Gnetum gnemon* (called *tulip* in Pidgin) fruit are totally accomplished with the labor input of an individual, the former by males and the latter mainly by females. Fishing can also be done by individuals, although at least two persons go together for any fishing expedition. The other activities are accomplished by division of labor and/or cooperation. Group hunting requires as many hunters as possible, and actually all adult males who are then staying in the village settlement participate. In contrast, horticulture, sago exploitation, coconut exploitation and *Canarium* fruit collecting involve both men and women. The division of labor by sex plays a basic role in these activities. In horticulture, cutting trees for new garden plots and making fences are men's tasks, while cutting shrubs and grasses in the new garden, weeding, planting and harvesting are women's. In sago-making, felling trees and removing the surface wood are men's tasks, while pounding and washing out pith are women's. In coconut-fetching and *Canarium*-collecting, men climb trees to throw down fruit and cut down trees, respectively, while women husk the fallen coconuts and pluck off the fallen *Canarium* fruits. These differences between male and female roles reveal that male tasks, more than female tasks, generally require physical strength and skill while women's work frequently requires repetition.

This contrast leads to the management of family labor which is essentially based on a sexual division of labor. In emergencies such as childbirth or illness, the husband is required to perform the women's tasks, especially in obtaining staple plant foods, while gifts from women of other households make a partial contribution. In fact, I observed married men making sago flour several times, and harvesting of garden crops is occasionally done by men, even on usual days. On the contrary, most male tasks are not customarily or cannot be done by women, but this does not usually interfere with daily subsistence activities. For example, cutting trees in new gardens needs several days' work during any ten-week period of the dry season, mainly during September and October. In cases where men cannot cut down sago palms or climb coconut trees, their wives join other sago-making or coconut-fetching parties to obtain the products. If a man cannot hunt for long time, meat from game animals obtained by individual hunting or group hunting is distributed fairly impartially to all households.

Three activities, i.e., sago exploitation, coconut exploitation and *Canarium* collecting, are usually organized by people from different households and membership varies from case to case; thus these activities present meaningful problems in labor organization. Of these three, *Canarium* collecting requires short work at an area near the settlement and is carried out by a group which is temporarily and very incidentally organized. Furthermore, *Canarium* trees grow naturally and are not looked after, while sago and coconut stands are owned by specific married males.

The adaptive function of labor organization is seen in sago exploitation and coconut exploitation. There were 32 cases of sago-making activities and 9 cases of coconut-fetching activities, in which group organizations were fully identified. None was performed by a single person. While some were accomplished by a single married couple, the working party usually consisted of from several to more than ten persons. In order to clarify the core cooperative relationship among all adult villagers, excluding *kewalbuga* males who participated in production parties very sporadically and arbitrarily, the frequency of every combination of two persons is analyzed as follows; for instance, the relation between two persons, A and B, is calculated by the formula:

$$\frac{1}{2}\times\left[\frac{\text{no. of cooperative efforts involving A and B}}{\text{total no. of cooperative efforts of A}} + \frac{\text{no. of cooperative efforts involving A and B}}{\text{total no. of cooperative efforts of B}}\right].$$

In this fashion, every combination is labeled by a numeral which is called here cooperative frequency, and a matrix of the cooperative frequencies for all individuals has been devised (Table 34). As a matter of course, when two persons worked together at any time and either of them did not work without the other, the cooperative frequency between the two is 1. This matrix includes all married villagers dwelling at the Ginar main settlement (including the dual dwellers of both Ginar and Gimurgi, namely Household Nos. 5 and 11) except for a *rugajog* man (Indiv. No. 4–1) who worked only a few times at sago-making and coconut-fetching in the intensive investigation periods for this analysis because of his temporary absence from the village on a trip to Daru.

Table 34. Cooperative frequency matrix in sago-making and coconut-fetching

1-2	2-1	2-2	3-1	3-2	4-2	4-3	5-1	5-2	6-1	6-2	6-3	7-1	7-2	8-1	8-2	9-1	9-2
1.00	0	0	0	0	0	0	0	0	0	0	0	0	0	0	0	0	0
	0	0	0	0	0	0	0	0	0	0	0	0	0	0	0	0	0
		1.00	0.17	0.17	0	0	0	0	0	0	0	0	0	0	0	0	0
			0.17	0.17	0	0	0	0	0	0	0	0	0	0	0	0	0
				1.00	0.17	0.12	0	0	0	0.15	0.12	0	0	0	0	0	0
					0.17	0.12	0	0	0	0.15	0.12	0	0	0	0	0	0
						0.44	0	0	0	0.18	0.45	0	0	0.35	0.35	0	0
							0	0	0	0.13	0.38	0.21	0.34	0.30	0.30	0	0
								1.00	0	0	0	0	0	0.50	0.50	0.42	0.50
									0	0	0	0	0	0.50	0.50	0.42	0.50
										0.83	0.53	0	0	0	0	0	0
											0.40	0	0	0	0	0	0
												0	0	0.30	0.30	0	0
													0.87	0	0	0	0
														0	0	0	0
															1.00	0.42	0.38
																0.42	0.38
																	0.87

[a] Except for Indiv. No. 4-1, for the reason mentioned in the text.

For extracting a simplified model of this matrix, a cluster analysis was applied. By this method, the two individuals whose cooperative frequency was the highest in the matrix were first clustered together, and then this new cluster was treated as a single unit, the same as an individual. For the reworking process, such a unit's cooperative frequency with each of the other individuals was defined as the mean value between the two values of the original cooperative frequency. This process was repeated to make one cluster involving all individuals. The result is illustrated in Figure 20, which excludes the clusterings beyond a frequency of 0.3. In this illustration, the solid line links individuals and/or groups interrelated by a frequency of more than 0.7, and the dotted line links those interrelated by a frequency between 0.3 and 0.7. Six out of the twelve monogamous couples worked together on any given occasion, and each of the other monogamous families as well as a husband and one of his wives in each of for Ginar-dwelling married people[a]

12-1	12-2	13-1	13-2	14-1	14-2	15-1	15-2	16-1	16-2	17-1	17-2	18-1	18-2	18-3	19-4	
0	0	0.58	0.57	0	0	0	0	0	0	0	0	0	0	0	0	1-1
0	0	0.58	0.57	0	0	0	0	0	0	0	0	0	0	0	0	1-2
0	0	0.18	0.17	0	0	0	0	0	0	0	0.15	0.27	0.37	0.60	0	2-1
0	0	0.18	0.17	0	0	0	0	0	0	0	0.15	0.27	0.37	0.60	0	2-2
0	0	0.15	0.29	0.14	0.13	0.13	0.13	0	0.13	0.34	0.36	0.24	0.31	0.17	0.17	3-1
0	0	0.15	0.29	0.14	0.13	0.13	0.13	0	0.13	0.34	0.36	0.24	0.31	0.17	0.17	3-2
0	0	0	0	0.69	0.65	0.65	0.62	0.17	0.33	0	0	0.27	0.37	0.20	0.60	4-2
0	0	0.13	0.12	0.47	0.54	0.54	0.61	0.82	0.86	0	0.19	0.21	0.26	0	0.58	4-3
0.42	0.42	0.33	0.32	0	0	0	0	0	0	0.35	0.30	0	0	0	0	5-1
0.42	0.42	0.33	0.32	0	0	0	0	0	0	0.35	0.30	0	0	0	0	5-2
0	0	0	0	0.31	0.29	0.29	0.28	0	0	0	0.13	0	0	0	0	6-1
0	0	0	0	0.31	0.29	0.29	0.28	0	0.15	0	0.13	0.25	0.17	0	0.18	6-2
0	0	0.13	0.12	0.49	0.56	0.45	0.52	0.36	0.34	0	0.10	0	0.13	0	0.15	6-3
0	0	0	0	0.24	0.23	0.23	0.22	0.24	0.23	0	0.22	0	0	0	0	7-1
0	0	0	0	0.20	0.19	0.19	0.18	0.20	0.19	0	0.18	0	0.21	0	0.23	7-2
0.42	0.42	0.33	0.32	0.32	0.31	0.31	0.31	0.32	0.31	0	0	0	0	0	0.35	8-1
0.42	0.42	0.33	0.32	0.32	0.31	0.31	0.31	0.32	0.31	0	0	0	0	0	0.35	8-2
0.33	0.33	0.50	0.48	0	0	0	0	0	0	0	0	0	0	0	0	9-1
0.29	0.29	0.42	0.39	0	0	0	0	0	0	0.23	0.18	0	0	0	0	9-2
	1.00	0.25	0.24	0	0	0	0	0	0	0	0	0	0	0	0.27	12-1
		0.25	0.24	0	0	0	0	0	0	0	0	0	0	0	0.27	12-2
			0.93	0	0	0	0	0	0	0	0.13	0	0	0.18	0.18	13-1
				0	0	0	0	0	0	0.17	0.24	0	0	0.17	0.17	13-2
					0.91	0.94	0.89	0.29	0.40	0	0.12	0.24	0.31	0	0.34	14-1
						0.88	0.94	0.40	0.50	0	0.23	0	0.29	0	0.33	14-2
							0.91	0.40	0.50	0	0.11	0.23	0.29	0	0.33	15-1
								0.51	0.59	0	0.21	0.22	0.28	0	0.31	15-2
									0.94	0	0.24	0	0	0	0.17	16-1
										0	0.23	0.23	0.21	0	0.33	16-2
											0.75	0	0	0	0	17-1
												0	0	0.15	0	17-2
													0.75	0.27	0.27	18-1
														0.37	0.37	18-2
															0.20	18-3

the polygynous families worked together at a frequency of more than 0.75 (one husband of the three polygynous families was Indiv. No. 4-1, who was excluded in this analysis because of the above-mentioned reason). These figures clearly imply that all monogamous couples form core units of cooperation, and in the polygynous families a husband tends to cooperate with one of his wives. This is reasonable simply because such core units equal meal units.

The solid lines, which combine persons other than married couples, indicate interesting links. The linking of four individuals (Indiv. Nos. 14-1, 14-2, 15-1 and 15-2) in two couples is first made through a two-male bond and a two-female bond. These two husbands and wives are respectively real brothers and sisters. Hence, this case suggests the strong cooperative bond of the brother-sister relationship. My second observation centers upon the link between a couple from Household No. 16 and Indiv. No. 4-3. Consanguineously, this

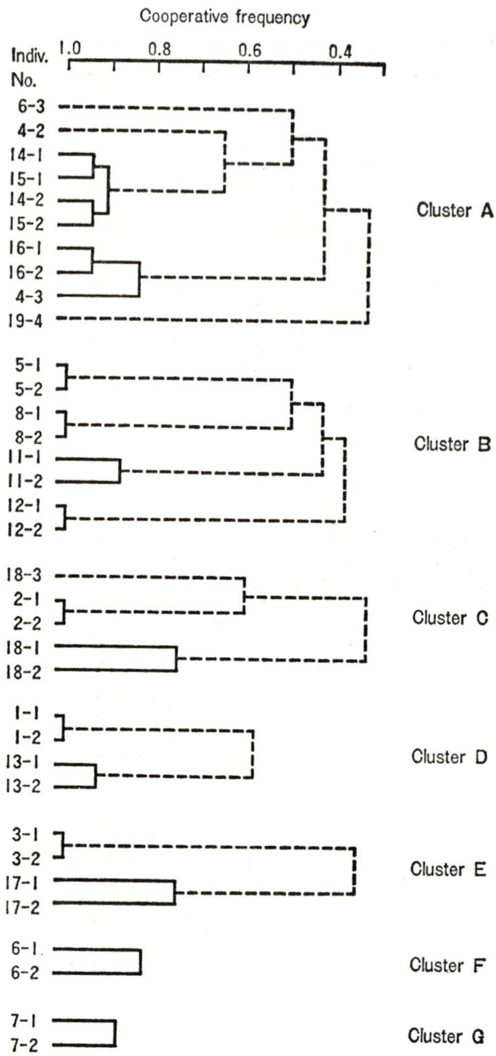

Fig. 20. Clustering of All Married Villagers by Cooperative Frequencies.

polygynous wife (Indiv. No. 4-3) is a sister of the husband of Household No. 16, so that this case also shows the importance of the brother-sister relationship.

Taking into account all cooperative links which have a frequency of more than 0.3, all individuals can be divided into seven clusters. These individual relationships state that most married couples have an expanded cooperative relationship with some specific couples and that any individual such as a widow and a polygynous wife, whose cooperative frequency with her husband is low, belongs to some cluster beyond the 0.3 average cooperative frequency. Since each of the seven clusters has its own particular structure, the following discussion treats them separately. In order to simplify the relationships among the people, the discussion will be restricted, as far as possible, to assessments based on parent-child and brother-sister relationships.

Cluster A. This cluster involves two core clusters, both of which are formed by married couples and brothers and sisters, as mentioned above. Furthermore, two polygynous wives (Indiv. Nos. 4-2 and 6-3), who are linked with the first core cluster at 0.65 and 0.50 levels, are sisters and also the sisters of the two husbands of Household Nos. 14 and 15. The linkage of the two core clusters at 0.43 level does not accord with parent-child nor brother-sister relationships, but the four siblings (Indiv. Nos. 6–3, 4–2, 14–1 and 15–1) in the first cluster and the two siblings (Indiv. Nos. 16–1 and 4–3) in the second belong to the same clan (Gameam clan) and their fathers are brothers. Only one widow (Indiv. No. 19–4) in the Wonie village, who has no brothers, sisters or children there, is clustered with this group, although the level of linkage is less than 0.4. The fact that the linkage of 0.4 is the highest among her cooperative frequencies reflects her membership in various working parties.

Cluster B. This cluster is composed of four core clusters, each being formed by a monogamous couple. The high linkage levels of these four contrasts with the lower levels of the larger clusters. It is impossible to elucidate consanguineously the formation of these large clusters, although these four couples have many of their sago and coconut groves in the northern part of the village-land, close to the Gimurugi small settlement, and thus the possibility of their cooperation is naturally high.

Cluster C. This cluster involves five people from two married couples, one monogamous and one polygynous, although a wife of the polygynous couple (Indiv. No. 18-3) first links not with a cluster involving her husband, but with another cluster of a couple of Household No. 2. The first linkage of the polygynous wife with the married couple of Household No. 2 cannot be fully related according to kin or affinity, but it is the case that this polygynous wife does not form a meal unit with her husband and has no close relatives in Wonie. The wife of Household No. 2 and the husband of Household No. 18 are brother and sister, and the two husbands were the exchange-partners of the wives at marriage.

Cluster D. The combination between two monogamous couples (Household Nos. 1 and 13), each forming a core cluster, is based on a parent-child relation. One of the female children of Indiv. Nos. 1-1 and 1-2 has married into Household No. 13.

Cluster E. The bond between the two monogamous couples does not reflect parent-child or brother-sister relations, but the wife of Household No. 17 and the husband of Household No. 3 are cousins, and according to the villagers the former was looked after by the latter during her unmarried period after her mother's death.

Clusters F and G. Each of the two clusters is composed of a couple and is characterized by a lack of clustering with other persons within the 0.3 level. Both heads are migrants to Wonie at or after the time of their marriages, Indiv. No. 6-1 from Podare and Indiv. No. 7-1 from Zim. The fact that Indiv. No. 6-1 joined the Wonie villagers more than twenty years ago indicates that village transference makes it difficult for the people to cooperate with their new village mates, even after a long time.

This cluster analysis has disclosed the principles of group organization for sago and coconut exploitation. The strongest bond is between married couples, while polygynous couples produce a core unit consisting of a husband and one of his wives, with the other wife associated with a different unit. The second principal bond is based on the brother-sister relationship when the siblings concerned include a male. The most conceivable reason for this is that sago and coconut stands, inherited from the siblings' father and owned by one male member among them, are to a considerable degree recognized as common properties. In other words, the ownership of these properties by a male appears to reflect his superintending them for the others. The third principle, although not to a similar extent, is associated with the parent-child relationship.

The organization of cooperative parties stabilizes the harvest of sago starch and coconuts for people like widows, polygynous wives and young married couples who possess few or no stands of these palms. Next, emphasis should be put on the cooperation of individuals beyond any close kin or affinal relations, even when this type of bond is less frequent than brother-sister-based or parent-child-based bonds. This is because such opportunistic cooperation contributes to the stabilization of food procurement by increasing the availability of these palms at the appropriate times. In other words, it is uncertain even for the owners of many sago and coconut stands whether the mature trees will be available at the times when they need the starch and fruit. Finally, the cooperation assures a food supply for all households in such emergencies as illness or a husband's stay away from the village.

Use of Space

The people of Wonie exploit their 70–80-square-km village-land to survive, although the degree of utilization varies to a great extent from portion to portion. In order to examine where the villagers exploit their food resources, the village-land is divided into meshes of 1 km × 1 km, with the village as the center. This mesh analysis is applied to the distribution of the producing areas among all Ginar-dwellers as well as to the frequency of their actual use for the two periods of the time-recording survey (26 days in total). However, the measures selected as producing means are sago groves in swampy land, mature coconut stands, and slash-and-burn gardens under use during the period from December 1971 to February 1972, when I systematically studied each household's gardens. Sago groves along creeks are excluded because of their negligible productivity, and number of matured sago stands has not been selected as a measure largely because of the difficulty in identifying which palms have attained maturity.

The number of actual uses of each mesh is counted for the four major activities, i.e., sago exploitation, coconut exploitation, horticulture and hunting. Collecting and fishing, both of which involved small numbers of actual performances, are not included. Exactly speaking, the use of meshes for the former three activities referred to the use of the producing plots in each mesh. For every hunting trip, only one mesh was identified as the hunting ground based on the hunter's judgment as to where he killed an animal or, if there was no kill, where he spent most of his time, although actually he would state the most remote area on his route, and he would have had chances to hunt on the way as well. In the process of counting the spatial use, no weighting for labor time or number of coworkers was applied.

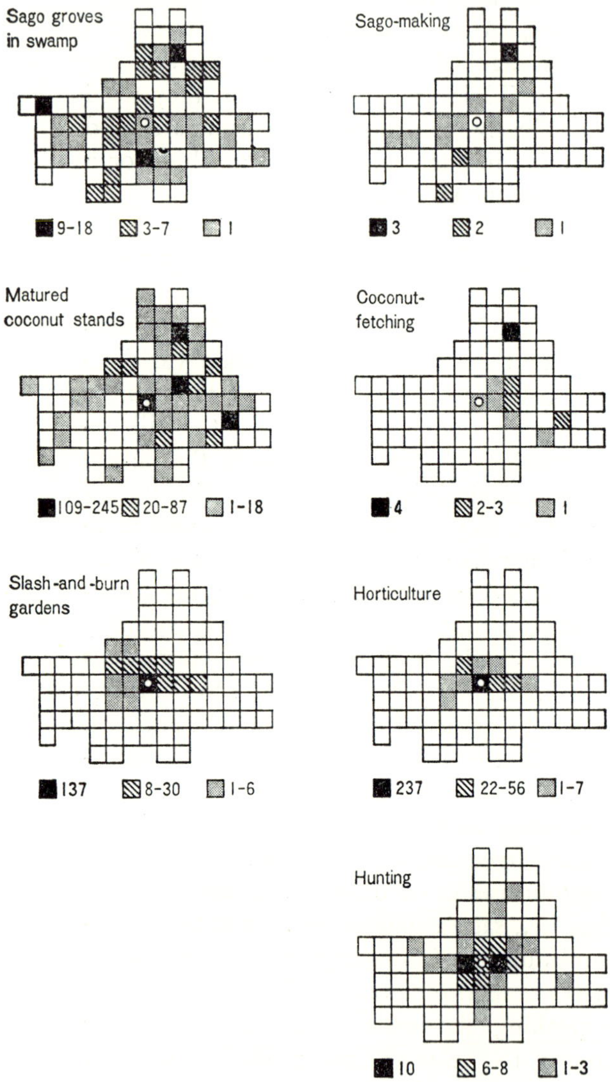

Fig. 21. Spaces Used as Producing Areas and for Food-Getting Activities on the Meshed Map of 1 km × 1 km.

The results are shown in Figure 21, in which the number of producing areas or usages is categorized into three, depending on the range of each figure. This analysis discloses three principal characteristics. First, the distribution pattern of producing areas for each activity pretty much parallels the pattern of actual use. The distribution of the actual uses of sago groves and coconut groves, especially that of the latter, appears to be biased by the distribution of the groves and stands, but this may owe greatly to contingency coupled with the paucity of my observations. Therefore, it is possible to consider the spatial utilization of these palms on the basis of the number of sago groves and coconut stands. Second, the distribution of sago and coconut groves is scattered over the village-land, in contrast to the gardens which are highly concentrated around the village. Finally, the distribution of meshes used in hunting is as wide as that in the exploitation of sago while a tendency towards concentration around the village, as in horticulture, is also seen.

The concentration of gardens in a spatial arrangement is the most extreme in the central mesh, which contains the village settlement. This mesh occupies 45 percent of the number of gardens and 65 percent of the used areas. Woodlands, which are suitable for gardens, thrive in savanna and along creeks in half the area of the village-land, and, according to my observations, the central mesh contained a wide area of such lands which still remained to be reclaimed. Thus, it can be concluded that horticulture requires about one square km and, even for a long period, one square km or a little more at most for the villagers as a whole.

On the other hand, there are two reasons for the fairly impartial distribution of sago groves and coconut groves throughout the village-land. The first reason is the long maturing time for these palms. Furthermore, since the palms grow naturally and the shoots of sago or the seedlings of coconut are frequently transplanted in the same grove as their mother palms, the groves tend to be utilized continuously, and the distribution of sago and coconut groves among contemporary dwellers reflects selections made by their ancestors long ago. As a natural consequence, the groves of these palms distribute over the village-land in its present stage.

The second reason applies only to the distribution of sago groves. Although coconut groves can be reclaimed in wide areas of savanna woodland or other vegetation, sago groves thrive exclusively in swampland. The land-specific nature of sago supports my observation that a number of small sago groves along creeks, made by the transplanting of shoots, provided very little sago starch, and most palm trees in such environments appeared to be less than fully mature. A possible cause comes from the complete drying up of such areas in the dry season of some years, although a phytophysiological assessment of sago's tolerance of dryness is not certain (Nishikawa, Oda and Suganuma 1979). The villagers told of extreme cases, where young sago stands in several such environments were burned during fires associated with group hunting and, more frequently, in other cases like maintaining the savanna environment. In spite of unsuitable environmental conditions, attempts at transplanting in creekside environments are frequently made. Thus, it is reasonable to conclude that all or most of the swamps suitable to sago palms in the village-land have long been used for sago groves. If we suppose that the density of sago is nearly at its maximum level, and that existing sago stands do not greatly exceed the minimum amount of sago starch which is required by this population, the village-land of 70–80 square km is necessary for sago exploitation, although the above supposition concerning the density of sago may not be applicable to some Gidra-speaking village groups along the larger rivers of the Oriomo, the Bituri and the Binaturi, because expansion of sago groves is highly possible there.

Another analysis relevant to the productivity of land is also necessary in terms of the distribution of hunting grounds. While the productivity of plant-food-getting activities is almost identical no matter where the producing areas are, hunting productivity may be influenced by the area utilized. For this purpose, all meshes are grouped into two zones, one nine square km consisting of nine meshes around and including the central mesh (called the near-zone) and the other involving the remainder (called the remote-zone). Table 35 shows total hunting time and total weight of game killed in each zone. In the table, all of the hunting time spent on a single activity is reflected in one mesh according to the hunter's

Table 35. Hunting time and game killed in the two zones. The classification of the two zones is mentioned in the text.

	near-zone	remote-zone
hunting time (hr)	127.6	219.8
game killed (kg)	166	350
game (kg/time [hr])	1.30	1.59

judgment of where he spent most of his time. As mentioned previously, such a place, in general, tends to be in the mesh most remote among those visited while hunting. Therefore, since hunting tours to the remote-zone allowed for additional hunting opportunities on the way, the actual ratio of time in the near-zone to that in the remote-zone would rise.

Even disregarding the above point, the importance of the remote-zone in hunting is emphasized not only by the two-thirds figure for total game animals obtained but also by its higher productivity per work hour. This difference between the zones perhaps comes from a decreased productivity for the land near the village owing to frequent disturbances of the animal's habitat by the people. Accordingly, it can be concluded that the nine square km of land around the village is insufficient for providing the people with game animals. Although there is no available information as to the animal population density, in order to avoid overexploitation the whole village-land may be needed for a continuous supply of animal meat over a long period. Along with the requirement of land for sago exploitation, this factor is perhaps a determinant of the human population density and/or population carrying capacity of land in the area.

2. Adaptation and Adaptability

The above section has documented and discussed the subsistence activity system of the Wonie villagers from various viewpoints. In the present study, the activity system is regarded as a key concept linking man with his environment, so that some further discussion is needed to synthesize and/or reassess these matters in the framework of man-environment relationships or adaptive mechanisms. Discussion will aim at extracting generalized ecological characteristics in human adaptation, as well as illuminating historical and comparative evaluations of the region-specific adaptive mechanisms. For this purpose, the various discussion points are grouped under three headings: "activity space as environment," "individuals and groups," and "dependence on sago and game." The former two belong to the man-environment category. The first heading pays particular attention to the environmental settings which immediately interact with human activity and human adaptation. Under the second heading, discussion is directed towards individual sociobiological nature such as sex and aging or life cycle, and the ecological implications of various human groups which are organized either formally or informally, permanently or temporarily. The last heading, "dependency on sago and game," is required because the core structure of the Oriomo Papuans' adaptive mechanism depends on the utilization of sago and game animals.

Activity Space as Environment

The findings of the present fieldwork stress the importance of the village-land not only as a natural environment *per se* but also as an activity space which is perceived, modified and actually utilized by the people. The degree to which the people utilize the village-land differs from section to section. On the other hand, the village-land is not used identically by all of the villagers. The idea of treating a living environment according to its actual use is fairly common, even if not systematically, among most field anthropologists and human ecologists and also parallels the opinions of some environmental scientists: for example, Kormondy (1974), who emphasized that all ecosystems in the world have been influenced by human activities and so they should be recognized not as natural ecosystems (or natural environments) but as "human ecosystems" (or environments artificially modified by human beings).

The Wonie villagers make minute distinctions within the village-land. This is typically reflected in the more than 200 place names in the 70–80 square km village-land. The villagers' identification of spots which have not yet been artificially modified includes not only micro-geographical and micro-vegetational information, but also the ever-changing conditions of such spots as they relate to hunting, collecting and fishing activities: for example, concerning individual hunting practices, whether a stand of trees whose fruit is a favorite of the cassowary bears fruit; how much water a waterpool still contains in the middle of the dry season and how often cassowaries have visited to bathe there; where the grasses which the grass-wallaby prefers grow densely; or where the mature wild tubers unearthed by pigs exist. For a group hunt, it is indispensable for the hunters to recognize whether grasses in the intended hunting ground have withered enough for their purpose. Also, the degree to which the creeks have been transformed into waterpools by the drying-up process during the dry season is carefully watched by the people in order to decide the best time for fishing by use of poison or with bows and arrows. The maturing process of edible wild fruits, the nesting of honey bees and so forth are always reviewed in their minds.

For plant-food-getting activities other than collecting wild plants, such minute observations of the natural environment play a lesser role. Instead, the people, especially adult males, are required to know the owner of each sago grove and coconut grove, and also the owner of each stand in the big groves which contain a number of stands. Such knowledge of possessions other than one's own is very useful for the mutual flow of information and thus for social solidarity. In other words, a man knowledgeable in these matters is recognized as "full-grown," and a *kewalbuga* male, who will enter the *rugajog* stage, is expected to be at this stage of knowledge.

These detailed perceptions of micro-geographical, micro-vegetational and faunal conditions in the village-land are essential for exploiting food resources in the local environment. It is the duty of the men to make these observations, and it takes a long time for *kewalbuga* to obtain the full amount of information. Because the information required is very specific and the region is narrow, however, this information cannot be directly applied anywhere except within their own village-land.

The manner of land exploitation greatly differs according to the kind of food-getting activity, as summarized in Table 36. One of the essential characteristics which is again emphasized

Table 36. Summarized characteristics of the major food-getting activities

	specification of environment exploited	work season	frequency or duration of use	ownership	inheritance
group hunting	savanna (involving patched woodland); specific	dry season	once a year at each spot	absent	—
individual hunting	any; depending on species to be hunted	any season	not specified	absent	—
sago exploitation	swamp (or creek-side)	any season	permanent (or semi-permanent)	married man	father to son
coconut exploitation	any drier land	any season	semi-permanent	married man	father to son
horticulture	woodland in savanna or along creek, etc.	labor cycle with season	several years	household	absent

is that sago palms and coconut palms grow permanently or semi-permanently in their respective groves, are each owned by a married male and are sometimes bequeathed through the male line from father to son. Thus, number and distribution of such producing areas are determined not only by the owner but also largely by his father and other ancestors; that is, they are mainly determined at birth. In other words, the activity space as environment differs from person to person, and the difference is largely a matter of inheritance. These factors should strengthen the male's adherence to his birth village, and this is reflected in the sexual difference in frequency of inter-village migration upon marriage. Among the contemporary Ginar-dwelling married members, three-fourths of the males but less than half of the females were born in Wonie. Of the four married men who were born outside the Wonie village-land, one (Indiv. No. 11-1) is a son of the Gimurugi-dwelling married couple, and his transference from Iamega to Wonie took place in his childhood together with his parents. Another (Indiv. No. 6-1), who is perhaps about sixty years old, has gradually extended his sago and coconut groves from the western end of the village-land, which is close to his birth village of Podare, to the center during the 20-year period of his stay in Wonie.

The other two household heads, i.e., Indiv. Nos. 7-1 and 17-1, are in their thirties and have stayed in Wonie for the past several years, the former having transferred from Zim and the latter from Gamaeve upon their marriages to Wonie women. Individual No. 7-1 does not yet own mature sago stands in the Wonie village-land, and therefore in order to obtain starch he, accompanied by his wife, frequently visits the village-land of Zim where he was born and owns sago groves. Individual No. 17-1's sago groves are concentrated close to the border of his birth village, Gamaeve, and within the Gamaeve village-land. It was revealed in my follow-up investigation in 1980 that the stay of these two *rugajog* men (and their families) in Wonie was not permanent but temporary because both had transferred to their respective birth villages before that time. In addition, the 1980 survey disclosed the return of the members of Household No. 8, whose head was born in Wonie and whose wife was born in Gamaeve, and both of whom had transferred to the wife's village during a later stage of my main survey in 1971 to 1972 (see pp. 47–50).

The final problem relating to activity space is how many people can survive on this village-land. The relation between population and the area inhabited is linked to a key term in ecology, "population density," as well as to another key term, "population carrying capacity of land" or, for short, "carrying capacity." In this study, taking into account considerations for hunter-gatherers (e.g., Casteel 1972–73), I will follow Allan's (1949) definition of this term for primitive agriculturalists, i.e., "the maximum number of people that a given land area will maintain in perpetuity under a given system of usage without land degradation setting in" because this definition may still be of high validity among anthropologists and geographers working in such communities (Street 1969). However, we must remember that the seemingly appropriate but puzzling term "carrying capacity," has caused various problems concerning the determination of the absolute volume of useful resources involved in the area. First, as emphasized by Hayden (1975), assessment over a long-term period is necessary to state the proper value. Second, it is reasonable to suppose that the density of population is certainly affected not by the area *per se* but by the quantities of specific food resources of essential importance. Another problem in the evaluation of carrying capacity stems from technological changes, which seriously alter the level (e.g., Boserup 1965:14; 1976).

My microdemographic analysis estimated the very slowly increasing population size in Oriomo around the last several decades (see pp. 20–21). The kinds of major food items and the technologies involved in food-getting activities among the contemporary villagers of Wonie do not appear to have changed much in the last several decades when compared to the ethnographic records of the early 1900's, as reported by Beaver (1920) and Williams (1936). Thus, it may be reasonable to judge that all three of the variables, consisting of the number of inhabitants, the kinds of food and the subsistence technologies in the area, have continued in a fairly stable state during the last several decades or the last century.

As has been discussed in the section "Use of Space," the area of swamplands suitable to sago palms and the density of game animals are two major factors which may condition the size of the human population in Wonie. Since virtually all swamplands in the village-land are already occupied by sago groves, the possibility of increasing sago groves is limited. The volume of game animals in the area was not determined nor estimated in my fieldwork, but the relatively higher yield and higher productivity in the zone remote from the village settlement is suggestive of a possible limiting factor for wild animals in this man-environment relationship. The importance of animal protein as a limiting factor in population density has been discussed in a comparative study among the aboriginal societies of the Amazon Basin (Gross 1975). Taking into account McCay's (1978) reasonable criticism of the premise about equilibrium between human population and resources in local environments, it can be tentatively concluded that in the subsistence system of the Wonie villagers, the quantities of sago and game availability are factors which keep the density of human population in the area at a level of less than one person per square km, possibly with a reservation for some access capacity, if not to the extent of the 60 or 70 per cent of carrying capacity that is probably reserved for stabilizing hunting-gathering populations (Lee and DeVore 1968), which parallels the underuse of resources as a component of Sahlins's hypothetical theory of pristine "affluence."

Individuals and Groups

For coping with the activity space mentioned above, the subsistence system of the people involves various sub-systems which are adjusted to the nature of the human population (or community), depending upon the various ages of individuals of both sexes, the formation of households and the replacement of members through generations. While all such sub-systems constitute parts of an adaptive mechanism, some of them are specifically concerned with individuals, some with households and some with the community as a whole. Discussion will begin with individuals.

The previous discussion about time spent in food-getting activities has disclosed that as people grow older, men's work changes qualitatively and quantitatively, while women's work changes only slightly in quantity. Male labor consists of hunting, cutting sago and big trees, climbing coconut trees and making garden fences. Generally speaking, this kind of work can be accomplished with physical strength and with skill which has been backed up by past experience. Hunting, especially, requires both traits, and the difference in hunting productivity among the groups according to age-grade is typical. Also, knowing when to cut sago palms is a matter of experience. In addition, the married males are obliged to familiarize themselves in detail with their village-land because the men make most critical judgments concerned with subsistence activities.

In adjusting to these conditions, the peculiar activity pattern of *kewalbuga* males is of special significance. Most of their time in actual subsistence activities is spent hunting. The hunting skill of the *kewalbuga* increases not only experience in shooting arrows *per se* but also knowledge of the animal ecology and the micro-environmental conditions of their activity space; in general terms, the importance of knowing animal ecology and environmental settings for hunting was emphasized by Laughlin (1968). The *kewalbuga* are not obliged to spend much time in plant-food-getting activities and so they have much time to acquire other necessary information and experience, such as micro-environmental knowledge, besides hunting techniques. Thus, the *kewalbuga* males have to depend on the married villagers for their daily food. The villagers themselves regard this age-grade as contributing to the sustenance of the villagers, not at the present, but in the future.

The age-grade system for males places the grade *kewalbuga* between the children's grade *yambuga* and the married men's grade *rugajog*, while for females the children's grade *ngamugaibuga* continues on to the married grade *kongajog*. In spite of this system, it appears that women who have just married and entered *kongajog* perform their duties without serious mishap. Another aspect of the establishment of the division of labor along sex lines lies in the different activity patterns in childhood. Boys, after they enter *yambuga*, tend to play separately from girls and sometimes follow *kewalbuga* males in travels outside the village. In such activities, bows and arrows are major playthings. On the other hand, as girls of *ngamugaibuga* grow up, they tend to observe adult women's labor although their direct contribution to subsistence production is negligible. Such a difference between boys and girls appears to satisfy sexual dimorphism in a biological sense and is also promoted by the discriminatory treatment of children by adult villagers, as is reported in detail by Mead (1975: 154–155) and Hogbin (1970) among other Papua New Guinean peoples.

Considering the fact that the sexual division of labor is a major strategy in their subsistence system, one may suppose that all food-getting activities can be completed by the labor force of a single household, or a husband and his wife or wives. This may be theoretically possible except in group hunting. Even if group hunting were not practiced, the males could obtain animals to a similar extent by individual hunting of higher productivity per work hour, although a great day-to-day fluctuation in animal yield is unavoidable. However, two necessary conditions are usually not met for the accomplishment of all subsistence activities within a single household. The first is a sufficient labor force. Aged men belonging to *nanyuruga* and *miid* tend to retire from hunting, cutting sago palms and climbing coconut palms and therefore cannot perform all of the male labors. The second condition is the lack of possession of a sufficient number of sago and coconut stands. In the household of a young husband and wife, the number of such stands is usually small, and the widow's household possesses no sago or coconut.

One or the other of these two types of handicap applies to many households. It may be more appropriate to say that all households are subject to both of the critical stages at some time in the life span. But they do not actually produce a serious problem because of the inter-household organization of such cooperative groups as sago-making parties and coconut-fetching parties, and the distribution system for game animals. The almost identical level of animal food and sago starch consumption among various households proves this (see Table 24). In other words, cooperation and/or mutual aid between households or at the community level are necessary for subsistence maintenance in all households.

It is also necessary to consider the compensations available when one of the spouses becomes seriously ill or a wife delivers a child. Men's labor, requiring physical strength, skill and decision-making, is seldom taken over by women, but if a husband cannot work even for several weeks his disability will not seriously damage his family's subsistence. There are three different reasons why these situations do not cause great problems. Garden-making, in which the men cut trees and make fences, can be performed during a period of several days at any time during a ten-week period in the dry season. Sago starch and coconut can be obtained by women's participation in work parties. Usually animal meat is impartially distributed among all households. In the latter two cases, systems of compensation are carried out at the inter-household or community level. In contrast, women's labor, which is characterized by its continuous nature, can be taken over by their spouses. In addition to the husband's participation, presents of food from various village members to the househod in question will normally occur.

Besides the above-mentioned roles for maintaining a stable supply of food for all households, there are two more community roles which assure the people's subsistence over time. One is the daily care of *kewalbuga* males by all married villagers who give them their food. This behavior assures the daily life of the *kewalbuga*, who are expected to learn hunting technique and to gain various knowledge for their future life as household heads. The other is the inheritance of sago and coconut palms. The married men own and care for the palms, not only for consumption but also for future use by their descendants. This system of inheritance appears to correspond to the fact that sago-making or coconut-fetching parties frequently involve brothers and sisters or children (and their spouses) of the owner of the palm to be utilized. Thus, the consanguineous and/or affinal relationships in the inter-generational framework assure the people's sustenance in spite of the absence of such formalized groups as extended families.

The above discussion has disregarded the dwellers of the Gimurugi small settlement, which consisted of two households (Nos. 9 and 10) having houses only there and two others (Nos. 5 and 11) having additional houses in the Ginar main settlement. Problems are found not only in the small size of the settlement, but also in the higher ages of the household heads, three (Indiv. Nos. 5-1, 9-1 and 10-1) of whom were *nanyuruga* and only one (Indiv. No. 11-1) of whom was a *rugajog*. These conditions might have produced an insufficient supply of animal food. Although I gathered no data for their food consumption and only sporadic data for their activities, their consumption of meat, which had to be supplied by their less-productive individual hunting practices and by presents from the Ginar-dwellers, might have been smaller in comparison with that of the Ginar-dwellers. At the time of my follow-up study in 1980, they had already abandoned the Gimurugi settlement; Household No. 9 joined the Iamega village, while the remainder settled down completely in the Ginar major settlement. This event is thought to reflect the disadvantage of a small settlement.

Dependency on Sago and Game

The core of the Oriomo Papuans' subsistence system depends on exploitation of sago simply because sago starch accounts for 60 percent of the total dietary energy consumption. In this final section, the region-specific adaptive mechanisms of sago-eating peoples in Melanesia or Oceania as a whole will be generally assessed in comparative ecological and prehistoric purviews while taking into account the findings of the present study and some available information from the literature.

The distribution of genus *Metroxylon* of sago covers most parts of Oceania, while in territories other than New Guinea and the neighboring islands, where varieties of *Metroxylon* exist, their use as food has been discontinued (Barrau 1961:60). When and why these plants dropped out of the food register in these areas still remains unknown, but, in Polynesia, Micronesia and eastern Melanesia, the *Metroxylon* species are plnated to provide thatching leaf for house-building, with the starch used only as an emergency food (Barrau 1959; Yen 1974; Ruddle, Johnson, Townsend and Rees 1978:66–69); it is of prehistoric significance that the original Proto-Melanesian word for sago palm means "thatch" (Chowning 1963). In spite of the above-mentioned facts, the value of starch extracted from the sago pith as a food has been, and still is, high for peoples inhabiting the low swamplands of Papua New Guinea, principally along the Sepik, Fly and Purari rivers and/or in several deltaic parts of the lowlands. The island of New Guinea still contains around 100,000 sago-eaters (McArthur 1972).

The plants listed as staple foods in New Guinea are, according to Powell (1976), coconut (*Cocos nucifera*), taro (*Colocasia esculenta*), yam (*Dioscorea alata* and *D. esculenta*), sweet potato (*Ipomea batatas*), sago (*Metroxylon* spp.) and banana (*Musa* spp.). Of these, only genus *Metroxylon* originated in the region (strictly speaking, perhaps in Indonesia), while the remainder are of Southeast Asian or American origins (Barrau 1968; Yen 1971; Powell 1976). However, the arrival of *Colocasia* in the region occurred very long ago and, according to recent archaeological evidence in the Highlands of New Guinea, a taro field dating back to 6,000 BP seems to have been identified (Golson 1977), while some wild and less-cultivated species of *Dioscorea* other than *D. alata* and *D. esculenta* might be of New Guinea origin. We do not yet have any archaeological evidence about *Metroxylon*, but there seems to be a strong possibility that this plant preceded taro and other tuberous crops in its arrival in New Guinea because of its indigenous nature.

If sago grew in Melanesia in such olden times, it might have grown completely wild. However, the "wild" versus "planted" difference in sago produces a variation in starch yield per tree (P. K. Townsend 1974), but the present study has proved that this difference does not greatly affect productivity per labor unit. One way to study the origin of sago exploitation is to assess the technology of sago-working. As implements used in sago-making throughout Oceania, steel axes have replaced stone tools for cutting palms almost everywhere, and in some places, e.g., among the Wogeo (Hogbin 1938–39) and the Tikopian (Firth 1939:135), the top of the sago-pounder, traditionally made of hard wood, bamboo or stone, has been fitted with a round iron placed at the striking edge. Other implements used in the general process are made of local materials and therefore are unlikely to have changed. The use of iron for the sago-pounder may not have changed labor efficiency very much because, despite the fact that the Oriomo Papuans know this new type of implement, most of them continue to use the traditional type. One might think that the steel axe is much more effective than the stone implement, but, according to W. H. Townsend (1969) who studied the Heve (Hiowe) living in the middle Sepik area, the sago-cutting stone implement used like an adze is about as effective as the steel axe for this purpose, and both tools were observed being used alternately to fell the same trunk in 1967.

To judge from the implements used for sago-making and the indigenous nature of sago in Melanesia, the invention of sago-making must have made possible the utilization of this productive food resource through all periods in the past. However, a more critical problem as yet unsolved is whether the invention of the complex sago-making technique, which includes pounding and washing out of the starch, was possible before shifting cultivation (cf. Barrau 1958:33–36; Nakao 1967:350). In contrast to the prevailing sago-making techniques throughout almost all sago-exploiting regions, there is an unusual method of extracting starch from sago pith reported by Wollaston (1912): "Very often the natives of the Mimika (in Irian Jaya) eat the crude sago, that is to say, the pith simply as it is cut out of the tree, without having been washed or pounded. The stuff is roasted in the usual way and the separation of the sago is done in the mouth of the eater, who spits out the uneatable fibre." Watanabe (pers. comm.; also see Golson 1972) observed that "this example suggests that wild sago-palms could have been utilized as food by pre-horticultural peoples even if they did not know the complicated technique of sago-processing." On this point it is necessary to investigate the extent to which this method of separation of sago pith is effective for the utilization of sago as a food, although the contemporary Wonie villagers do not know this method, even in legend, and believe it impractical.

Apart from a historical evaluation of sago-making technology, sago should be compared with other staple food plants in the subsistence system. Comparing the number of man-hours needed to produce caloric return among several different economies, P. K. Townsend (1974) places sago production among the Sanio-Hiowe in the Sepik area, which is almost identical to that of the Oriomo, on a comparable level to swidden cultivations in other societies. Nonetheless, among the Oriomo Papuans the labor efficiency of sago production is superior to that of garden cultivation (see section on Labor Efficiency). This is mostly due to the fact that little labor other than harvesting is required for the utilization of sago. In addition to sago's advantage in labor productivity, it is also characterized by its year-round availability, unlike most wild roots, fruits and some garden crops. Moreover, sago tends to be little harmed by wild animals such as pigs and rats or by climatic abnormalities. These characteristics of sago greatly contribute to its stability as a dietary source. The potential storage period for sago starch is fairly long, although in many societies where sago starch is utilized as the most important food it is usually consumed within a couple of weeks after production.

Along with the advantages, a system of subsistence based on sago also has some disadvantages. The system depends upon a very specific habitat, namely low swampy land, and is restricted to this environment. One potential advantage in transplanting sago is the expansion of the area in which sago palms grow. However, the Wonie villagers' attempts to transplant sago shoots in places along a number of small creeks have met with very limited success. These new groves contain only a small number of full-grown sago stands and yield little starch. For people who depend on wild sago palms, the density of palms on their land is one of the main factors determining the human population density, and even for people who depend on planted sago palms it is impossible to raise the density of palms beyond the capacity of the swampy area. This leads to two conclusions. First, it is impossible for people who depend on sago, wild or planted, to migrate out of the swampy land unless they change their staple food as well as their total subsistence system and adaptive mechanisms. Second, their human population density, which is conditioned by ecological and geographical restrictions, would seem automatically limited.

In such societies as the Oriomo where sago palms are owned by particular persons and inherited through generations, the people's adherence to their native place is incorporated in the adaptive mechanism. The findings of my fieldwork support the conclusion that the major sago groves of each villager are distributed in the area close to his ancestors' central activity place (see Fig. 14). Two married men (Indiv. Nos. 6-1 and 7-1) who migrated to Wonie from adjacent villages own no inherited sago there, and the younger of the two still has to make sago starch completely outside the Wonie village-land; two married Wonie men (Indiv. Nos. 7-1 and 17-1) in their thirties, who had been born in other villages, transferred to their respective birth villages before my follow-up study in 1980, while, on the contrary, the Wonie man (Indiv. No. 8-1) who had transferred, with his wife, to his wife's village in my main survey period came back to Wonie before 1980.

A further disadvantage which stems from the geographical restriction is that the swampland where sago-eaters live is generally unfavorable to health. The proportion of stillbirths plus infant deaths to total births, which occurred during my stay in Wonie, was about half, and although no medically accurate information was obtained, it seemed to me that some of the surviving babies looked unhealthy and suffered at least from serious skin diseases.

It should also be emphasized that nutritionally speaking, sago starch contains little value other than carbohydrate. In comparison with other staples, tubers or cereals, sago is extreme in this regard. For this reason, this staple specifically requires dietary supplements, especially animal foods rich in protein. Unless domestic animals are available on a daily or semi-daily basis, natural resources must be exploited by hunting and/or fishing. Alternately, some researchers (e.g., Fountain 1966) have stressed the importance of sago grubs (family Cerambycidae) as a protein source, while Corden (1970) listed the leaves and fruit of *Gnetum gnemon* (called *tulip* in Pidgin) as a protein supply. In my view, however, these protein sources may not be commonly utilized to an extent which sufficiently satisfies protein demand.

Generally speaking, sago-eating people inhabiting coastal locations depend on fishing, while those inhabiting inland areas tend to hunt. The population density of the former may depend to a lesser degree on the environment, since marine resources are less likely to be affected by human exploitation than land animals, although most sago-eating populations live in the inland because sago thrives in freshwater swamps. On the other hand, the population density of the inland areas may depend not only on the density of sago palms but also on the amount of wild animal meat available. Thus, this type of adaptation has proven vulnerable to demographic and social change. A typical case among the sago-eating Yimar of the upper Korobori River was reported by Haberland and Seyfarth (1974:243): "Since World War II, as a consequence of acculturation, they settle in larger, composite villages and have abandoned hunting without providing an adequate protein replacement." This case of the Yimar is very suggestive of an ecological condition in the adaptation of sago-eaters. In fact, the Oriomo Papuans are one of the sago-eating populations that depend very heavily on hunting. The availability of game animals should determine the balance between animals and people in the local environment and thus account for the low human population density.

P. K. Townsend (1971, 1974) points out another nutritional problem. Sago is not suitable as a food for infants and therefore breastfeeding is necessary for a long period after birth. She further supposes that this drawback of sago is one of the causes for persistent infanticide among the Sanio-Hiowe, although I obtained no information about infanticide among the contemporary Oriomo Papuans. Parallel claims of indigestibility were reported among several peoples; for example, the Toraja of Sulawesi forbid consumption of sago by persons with abdominal pain or swellings, believing that sago may contribute to abdominal swelling (Ruddle, Johnson, Townsend and Rees 1978:61).

The above discussion of the existing field research and information from other societies has clarified a generalized pattern of subsistence and adaptation among sago-eaters. The system of subsistence which depends on sago is stable not only throughout the year but also over periods of years. Whatever the reason, the low density of sago palms, the low density of wild animals, the unfavorable conditions for health, or the poor nutritional levels of sago, the population density of human groups heavily dependent on sago is low. Also, these factors affect the survival of the people, especially infants, on the one hand and, on the other, inhibit them from elaborating their socioeconomic or sociopolitical organizations.

Epilogue

In my understanding, the primary efforts of human ecologists and/or ecological anthropologists are to tap adaptive mechanisms of various human populations who survive in local ecosystems. For this purpose, we must first simultaneously utilize a variety of field methodologies involving both the so-called natural sciences and the social or human sciences. Second, some aspects of the study are to be focused upon individuals, and some upon small groups such as households, families and communities, while ideally or conceptually the human ecology-specific approach should deal with human aggregations as a *population* defined by ecologists and geneticists, like "a collective group of organisms of the same species occupying a particular space" (Odum 1971: 162) and/or "a reproductive community (biotic community) of sexual and cross-fertilizing individuals which share in a common gene group" (Dobzhansky 1968). The Gidra-speaking group is one of the best-suited to this approach because, in spite of its small size, this group has possessed a specific habitat and has continued intermarriage.

Practically, we have two rather contradictory aims for the research strategy. One is designed for investigating the subjects in as much detail as possible, while the other is designed for investigating the entire subject group or as many members as possible. My own fieldwork among a single village, upon which the present monograph is based, clearly aimed at the first strategy. For satisfying the above two points, especially the second point, a new project based on the group efforts of five Japanese and one Papua New Guinean researchers for the whole Gidra-speaking population began in 1980 and is ongoing. According to the field experience and raw data of the new project, one of our achievements will be to show some of the significant problems in the holistic dynamics of the whole human population. Another meaningful achievement will be to reiterate the significance of intensive investigation in each village, because population ecological assessment is largely validated by comparative analyses of adaptive systems in various villages (sub-groups of the population) which vary among each other in natural as well as sociocultural environmental conditions.

Finally, the developmental-ecological assessment of a subsistence system, as emphasized, for example, in Papua New Guinean peoples by Clarke (1976) and Bayliss-Smith (1977), may be a theme relevant to human population ecology. This is beyond the direct aim of the present monograph, but the above-mentioned findings derived from an intensive study in one village, which will possibly be coupled with the forthcoming results of our new project, may provide essential information for regional and community planning.

Appendices

Appendix 1. Tree species of the Morehead-Kiunga region.

Flora in the Morehead-Kiunga region, which includes a part of the Oriomo Plateau, around 100 km west of my study region, was studied in 179 one-third-acre sample plots by Paijmans (1971). The species of trees and their densities in each vegetation type seem to be fairly identical throughout the plateau. The trees in monsoon forests (which correspond with *bua*, the local term in my study area) are shown below. According to Paijmans's classification, the occurrence of trees is shown as D (dominant, 50–80%); SD (subdominant, 20–50%);

botanical name	commonness of occurrence	botanical name	commonness of occurrence
Acasia aulacocarpa	C	*Gmelina*	R
Anisoptera kostermansiana	R	*Hopea* 1	R (SD)
Buchanania	R	*Hopea* 2	R
Calophyllum	R (O)	*Lithocarpus*	R
Canarium	R	*Litsea*	R
Castanopsis acuminatissima	R	*Mangifera*	O
Cryptocarya	R	*Oreocallis brachycarpa*	R
Dillenia	R	*Parinari nonda*	O
Elaeocarpus	R	*Schizomeria*	O
Endospermum	R	*Sloanea*	R
Euodia	R	*?Sloanea*	R
Ficus	R	*?Stenocarpus*	R
Flindersia	O	*Syzygium*	O
Ganophyllum falcatum	R	*Terminalia*	R
Garcinia	R	*Vatica papuana*	R
Gironniera	R		

VC (very common; 15–20%); C (common; 10–15%); O (occasional, 5–10%); R (rare, <5%); (   ) (locally), although the categories of D and VC do not appear in this vegetation type. In the savanna (called *yap*, in my study area), some of these tree species grow as well, while such species as *Melaleuca* spp. occur extensively.

Appendix 2. A mythical story of the origin of the Dorem clan.

The first man of the Dorem, called Kuub,[1] was created along with a grass-wallaby (called *seba* locally) at Bier, about four km west of the present site of Wonie village. This founding ancestor was not a real man but something like a pole which stood at Bier. One day, a man of the Dumaram clan[2] spoke to Kuub, but he was able to move his body only slightly. The Dumaram man felt compassion for his gesture, and carved a face, hands, legs and other parts of his body with a knife.[3] Not only Kuub but also the first men of three other clans, Wammopam, Morob and Kameam, were created at nearly the same time at Wammopam, Unja and Kuma, respectively. One day these first four men set out together on a journey.

The first day of the journey.

The four ancestors traveled eastwards up along a small creek named Kobe. On the way, Morob parted from the other three to turn to the right. When the three men arrived at Goara, about seven or eight km east of Bier, they met two people, Masam and Idara, who might have been created separately at Masa and Idara near Goara, and came separately down along the same creek. As soon as the three people got to Goara,

[1] My informants do not know the names of the first men except their own, Kuub, because no one should and no one does know the names of the first men of any clans other than his own. For this reason, each of the first men in the other clans is represented by his clan's name itself in the story.

[2] According to the story, at the creation of their world only the people of Dumaram and Midam were real men, although both are recognized as ordinary clans at the present time.

[3] At that time *seba*, which is the *kakiena* (main totem) of the Dorem clan, may have also been created, although my informants do not remember accurately by whom and from what it was made.

Idara left for his home. After a while, all four ancestors started for their own homes separately.

The second day of the journey.

Kuub himself started northeastwards for Yokunja, taking with him legs of *seba* and testicles of *ruei* (a kind of bandicoot). On the way, he met Mognayam at Wumu about two or three km from Bier. Mognayam was also created together with *seba* in the same way as the creation of Kuub, at Nguya about three or four km northeast of Bier. When the two ancestors came across each other, Mognayam said, "We are like brothers born from the same father and mother." Kuub continued his journey for Yokunja with Mognayam. At Murenakagnemuneni, half way to Yokunja, Mognayam parted from Kuub to go northwestwards to Wanja, saying, "Goodbye." After that Kuub set out again for Yokunja himself.

At Yokunja, he met three persons, Yokunjam, Sebaiena and Tonyukakam, who lived separately at the three dents of a big, folded tree called Dom. Kuub presented *seba* legs to Sebaiena and said, "Don't eat these legs, only keep them." And he stayed that night with Sebaiena. Kuub, however, did not present anything to the other two persons nor talk to them.

The third day of the journey.

In the morning, Kuub set out by himself northwards for Daaru to continue his journey. On the way, he met Damuram at Wobor. Though he did not stop walking, he talked with him for a short time. Damuram said, "Who are you?" and Kuub answered, "Our clan is Dorem." So Damuram said again, "Our clan is Damuram." After that brief conversation, Kuub walked on northwards by himself. He met Ijobora at Iogor and they had exactly the same conversation as that between Kuub and Damuram. At last he arrived at Daaru, the destination of that day, where Daream lived. Daream said to Kuub, "We also belong to the Dorem," and Kuub presented testicles of *ruei* (bandicoot) to Daream, saying, "I will present these testicles of bandicoot to you, but please never eat them, only keep them." That night Kuub stayed with Daream.

The story ends at this point in time, that is, in the evening of the third day of their ancestor's journey. As far as the origin myth of the Dorem clan is concerned, classification of clans into two moieties is related to the story. The clans whose ancestors talked or walked with Kuub during his journey belong to the same moiety as the Dorem. Of the twelve clans which appeared in this origin myth, only three, i.e., Idara, Yokunjam and Tonyukakam, belong to a moiety other than that of the Dorem, and the remainder belong to the Dorem moiety, although there are more than thirty known clans in the Gidra linguistic group.

Appendix 3. Trace metals in Oriomo Papuans' hair.

Scalp hair samples of 39 Wonie villagers (one *sobijogbuga*, one *yambuga*, five *kewalbuga*, four *rugajog*, six *sobijogngamugai*, nine *ngamugaibuga*, twelve *kongajog* and one *nanyukonga*) were analyzed for concentrations of zinc, copper, and inorganic and organic mercury. A part of the results of the analysis is shown in the table below (see Ohtsuka and Suzuki 1978 for details).

Trace metal	Age-grade[a] (No.)	Mean	SD
Zinc (μg/g)	boys (2)	102	—
	men (9)	129	30
	girls (15)	93	34
	women (13)	93	25
Copper (μg/g)	boys (2)	8.6	—
	men (9)	11.8	1.8
	girls (15)	9.5	1.2
	women (13)	11.1	6.2
Inorganic mercury (μg/g)	boys (2)	0.22	—
	men (8)	0.24	0.11
	girls (15)	0.29	0.17
	women (13)	0.25	0.13
Organic mercury (μg/g)	boys (2)	0.85	—
	men (8)	1.54	0.38
	girls (15)	1.03	0.67
	women (13)	1.10	0.43

[a] Boys: *sobijogbuga* and *yambuga*. Men: *kewalbuga* and *rugajog*. Girls: *sobijogngamugai* and *ngamugaibuga*. Women: *kongajog* and *nanyukonga*.

The most prominent feature is a low level of zinc concentration in females; this is reconfirmed by another analysis using the same samples (Sasaki, Takemori, Ohtsuka and

Suzuki 1981). Compared to the report of Hambridge, Hambridge, Jacobs and Baum (1972), the zinc status of the Oriomo female may be deficient or marginal. Sever (1975) suggests that maternal zinc deficiency may contribute to the high incidence of human central nervous system malformations in some populations. Although information about malformations was not obtained among the Oriomo, there remains a possibility that the poor maternal zinc nutriture relates to the high incidence of stillbirths observed.

Appendix 4. Fishing at a sacred place, Goara.

A small area of about 1,000 m by 400 m, which is called Goara as a local place name, is the only sacred place throughout the Wonie village-land. This area forms a big depression, being a swamp in the wet season and a sterile land with only short grasses and a few trees in the peripheral zone in the dry season. According to their mythical story, a big *iit* (*kewalbuga*'s house) stood there in olden times. Some women happened to come close to the *iit*, violating the rule that women and boys are strictly required to keep away from it. At that moment, thunder and an earthquake occurred, killing the women, destroying all buildings and making the land a sterile depression. Since then, women and boys have been prohibited from coming into Goara, and men have also been prohibited from sleeping, excreting feces and urine, or cutting trees and grasses there.

Fish aggregate in the waterpool of Goara in the wet season, and as the dry season advances they are caught in the shrinking waterpool. The married and *kewalbuga* males are allowed to catch the fish there only once a year, a successive couple of days. During my study period for the 1971 dry season, the fishing took place in early September. For several days before 5 September, a number of *kewalbuga* and married males by twos or threes frequently made reconnaissance trips to Goara, one hour way from the village. According to the final judgment of a *miid* man (Indiv. No. 1-1) as to the right time for fishing, many married and *kewalbuga* males of Wonie moved to Goara on the afternoon of 6 September, accompanied by two *rugajog* men from the Kuru village and several *kewalbuga* males from the Iamega and Gamaeve villages. On that day, what the attendants did consisted of looking into the waterpool and constructing several huts in the area about 500 m apart from Goara.

On the following morning of 7 September, only several people went fishing using a rod-line, while some went hunting and some went back to the village to announce the proceeding and to fetch vegetable foods, and some others chatted in the huts. About an hour's fishing in the early morning caught some sixty fish, which were cooked and eaten by groups at sporadic intervals in the afternoon. The last adult man of Wonie arrived at Goara by noon. In the afternoon, the younger members collected a huge amount of firewood, wood for lighting, and wild palm fronds to serve as containers.

At 4:20 on 8 September, the members started the Goara-specific fishing. The method employed was exclusively bow-and-arrow shooting, and the archers used lightening wood, about 1.5 m long, before daybreak. Each archer stood on the shore one or two m apart from the water, holding a bow and arrows with one hand and a piece of lighting wood with the other. When he noticed a fish coming up to the surface (perhaps this occurred owing to the brightness of the light), he put down the light and shot an arrow. The drawing of the bow was so slight and so quick as to require only a couple of seconds between recognition and shooting. Two *rugajog* men were prohibited from actual fishing because of their wives' pregnancy, and several aged men spent much time at the camping site.

By 7:00, about twenty fishermen had caught more than 500 fish, the bulk of which were of four species and were 15 to 30 cm in length. After that, everyone left the waterpool for the camping site to cook the catch. They enjoyed the food while chatting for several hours. In the afternoon, they returned to the village by twos or threes, carrying the rest of the cooked fish. The fish can be eaten by women and children outside the sacred area, and actually many of them waited for the fish on the way to the village.

Appendix 5. An episode of women's hunting.

While Morren (1975:285–286) reports fairly popular practices of women's hunting among the Miyanmin of the Highlands of New Guinea, I observed Wonie women obtain land mammals only once throughout my study period. This occurred in the dry season. While on their way back from the garden three women accompanied by their children rested at the side of a path around four km east of the village. One of them found rats hidden in a hole of a blighted tree about two m above the ground. The women gathered a large amount of withered grass and leaves and pushed them into another hole just above the ground, and then set fire to them. After about ten minutes of smoke, an adult rat around one kg in weight and three juveniles were driven out of a different hole three m above the ground and were beaten with logs.

Appendix 6. Slash-and-burn gardens possessed by each household: year of first clearance and kinds of plants.

The observations were made at different times between December 1971 and February 1972; and consequently, not all new gardens were planted over. In the following list, *pitpit* (in Pidgin), pumpkin, watermelon, and cowpea are excluded because of their small amounts and poor growth. Abbreviations for plants: Ab=*Abelmoschus manihot,* or *abika* in Pidgin, Am=*Amorphophallus campanulatus*, Ba=banana, Ca=cassava, Pa=papaya, Pi=pineapple, Pn=pandanus, Sc=sugar cane, Sp=sweet potato, Ta=taro, and Ya=yam.

* Ill-kept garden which may be unproductive, although some other gardens without asterisks, especially those cleared before 1970, also involve only ill-grown plants such as pineapple, which may not be utilized even for transplanting shoots.

Household No. 1

year	kinds of plants
1968*	Ba, Pa
69	Ba, Pa, Pi
69	Pi
69*	Ba
70	Ta, Ba, Pi, Ca
70*	Ba, Pa
70*	Ba, Pa
70*	Ba
70*	Ba
71	Ta, Ba, Ya
71	Ta, Ba, Pi
71	Ta
71	Ta
72	Ta, Ba, Ya, Sc, Pn
72	Ta, Ba, Am
72	Ba

Household No. 2

year	kinds of plants
1968*	Ba, Pa
69	Ba, Pa, Pi
69*	Pi
70	Ba, Pi
70*	Ba
70*	Ba
71	Ta, Ba, Pa, Pi, Sp
71	Ta
71	Ya
72	Ta, Ba, Am, Ya, Ab
72	Ta
72	Sp, Ya
72	Ba, Pi

Household No. 3

year	kinds of plants
1968*	Ba, Pa
69	Ba, Pa, Pi
69*	Ba
69*	Pi
70	Ba, Pi
70*	Ba
70*	Ba
71	Ta, Ba, Pa, Pi, Sp, Pn
71	Ta
71	Ta
71	Ya
72	Ta, Am, Ya, Sc
1972	Ba, Pi, Pn
72	Ba, Pa, Pi
72	Ba, Pi
72	Pi, Sp, Ca

Household No. 4

year	kinds of plants
1968*	Ba, Pa
69	Ba, Pa, Pi
70*	Ba
70*	Ba
71	Ta, Ba, Pa, Pi, Sp, Pn
71	Ta, Ba, Am, Ya, Ca, Sc
71	Ta, Sc
72	Ta, Ba, Ya, Ca, Sc
72	Ta, Sc
72	Ba, Pi, Sc
72	Ba, Pi

Household No. 5

year	kinds of plants
1969	Ba, Pa, Pi, Sp, Sc
69	Ba, Pa, Pi
69	Ba, Pa, Pi, Sc
70	Ba, Pa, Pi, Pn
70*	Ba
71	Ta, Ba, Pi, Sp, Ya, Ca, Sc
71	Ya
72	Ta, Am
72	Ba, Pi
72	Pi

Household No. 6

year	kinds of plants
1968*	Ba
69	Ba, Pa, Pi
69*	Ba, Pa
69*	Ba, Pi
69*	Ba
70	Ta, Ba, Pi
70	Ba, Pi
70*	Ba
70*	Ba
70*	Pi
71	Ta, Ba, Ya, Sc
71	Ta, Ba, Pi, Pn
71	Ta, Sc
71	Ba, Pi
72	Ta, Am, Ya

1972	Ba, Pi, Sc
72	Ba, Ca

Household No. 7

year	kinds of plants
1969	Ba, Pa, Pi
69*	Pi
70	Ta, Ba, Pa, Pi, Ca
70	Ba, Pa, Pi
71	Ta, Ba, Pa, Pi, Sp, Pn
71	Ta, Sc
71	Ya
72	Ta, Ba, Pi, Am, Pn
72	Ta, Ba, Pi
72	Ta, Ba

Household No. 8

year	kinds of plants
1970	Ba, Pa, Pi, Sp
70	Ba
71	Ta, Ba, Pa, Pi, Ca, Pn
71	Ta, Ba, Pa, Pi, Ca, Pn
71	Ya
72	Ta, Ba, Pi, Am, Ab

Household No. 9

year	kinds of plants
1969	Ba, Pa, Pi, Sc
69	Ba, Pa, Pi
70	Ba, Pa, Pi, Pn
70	Ba, Pa, Pi
71	Ta, Ba, Pa, Pi, Sp, Ca, Sc
71	Ya
72	Ta, Ba, Am, Pi, Ya, Sc, Pn
72	Ta, Ba, Am
72	Ba, Pi
72	Ya

Household No. 10

year	kinds of plants
1969	Ba, Pa, Pi, Sc
69	Ba, Pa, Pi
70	Ba, Pa, Pi, Pn
70	Ba, Pa, Pi
71	Ta, Ba, Pa, Pi, Sp, Ca, Sc
72	Ta, Am, Ya, Sc, Pn
72	Ta

Household No. 11

year	kinds of plants
1968*	Ba
69	Ba, Pa, Pi, Sc
69*	Ba, Pa
70	Ba, Pa, Pi
71	Ta, Ba, Pa, Pi, Sp, Ca, Sc, Pn
71	Ta, Sc
71	Ya
72	Ta, Am, Ya, Sc
72	Ba
72	Ba
72	Pi

Household No. 12

year	kinds of plants
1968*	Ba
69	Ba, Pa, Pi, Ca, Sc, Sp
69*	Ba, Pa, Pi
69*	Ba
70	Ta, Ba, Pi, Ca, Pn
70	Ba, Pa, Pi
70	Ba, Pi
71	Ta, Ba, Pi, Sp, Ya, Ca, Sc
71	Ta, Ba, Pi
71	Ya
72	Ta, Ba, Am
72	Ta, Ba, Pi, Pn

Household No. 13

year	kinds of plants
1968*	Ba
69	Ba, Pa, Pi, Sp, Pn
70	Ta, Ba, Pa, Pi, Ca
70	Ba, Pa, Pi
70*	Ba
71	Ta
71	Ba, Pi
71	Ya
72	Ta, Ba, Am
72	Ta
72	Pi

Household No. 14

year	kinds of plants
1968*	Ba
69	Ba, Pa, Pi
69*	Ba

1969*	Ba
70*	Ba
71	Ta, Ba, Pa, Pi, Sp, Ya, Sc
71	Ta, Ba, Pa, Pi, Sc, Pn
71	Ta, Sc
72	Ta, Ba, Am, Ya, Sc
72	Ta, Ba, Pi, Sc
72	Ba, Pa, Pi
72	Ba, Pi

Household No. 15

year	kinds of plants
1968*	Ba
69*	Pa, Pi
69*	Ba
70	Ba
71	Ta, Ba, Pa, Pi, Ya, Sp, Sc
71	Ta
72	Ta, Pi
72	Ta, Ba, Am
72	Ta
72	Ba, Pi
72	Ba

Household No. 16

year	kinds of plants
1968*	Ba
69	Ba, Pa, Pi
70	Ba
71	Ta, Ba, Pa, Pi, Ya, Sp, Sc, Pn
71	Ta
71	Ba, Pi
72	Ta, Ba, Am
72	Ta, Am, Sc
72	Ta
72	Ba

Household No. 17

year	kinds of plants
1968*	Ba
69	Ba, Pi, Pn
70	Ta, Ba, Pi, Ca, Pn
70	Ba, Pi, Sp, Sc
70	Ba, Pa, Pn
70*	Ba
70*	Ba
70*	Pi
71	Ta, Ba, Pa, Ya, Sc
71	Ta
71	Ba, Pi, Sc
71	Ba, Pi
72	Ta, Ba, Pa, Pi, Am, Sc, Pn
72	Ba, Pi, Sc
72	Ba, Pi,

Household No. 18

year	kinds of plants
1968*	Ba, Pa
69	Ba, Pa, Pi
70*	Ba
70*	Ba
71	Ta, Ba, Pa, Pi, Sp, Pn
71	Ta
71	Ya
71	Ba, Pi
72	Ta, Ba, Am, Ya, Ca, Ab
72	Ta, Ba, Pi

Household No. 19

year	kinds of plants
1968*	Ba
69	Ba, Pa, Pi, Sp, Pn
70	Ba
70*	Ba
71	Ta, Ba, Sp
71	Ta, Sc

Appendix 7. Man-value coefficients used for adjusting person-to-person differences in dietary consumption.

This adjustment system was originally used by Thomson (1954).

adult man: 1.0
adult woman: 0.85
adult woman in pregnancy: 0.95
boy over 14 years old: 0.85–1.00*
girl over 14 years old: 0.80–0.85*
younger children of both sexes:

12–14 in age	0.80
11–12	0.75*
9–11	0.65
7–9	0.55
5–7	0.45
3–5	0.40
2–3	0.35*
1–2	0.25*
0–1	0.20*

* These categories were not applied since no subject person belonged to the age groups in question.

Bibliography

Akimichi, T., 1975, Individual tuna trolling strategies and transmission of fishing skill in a local community of Shimokita Peninsula. *Journal of Human Ergology*, 4: 83–101.

Allan, W., 1949, Studies in African land usage in northern Rhodesia. *Rhodes Livingstone Papers*, 15.

Altmann, S. A. and Altmann, J., 1970, *Baboon Ecology*. S. Karger, Basel.

Archbold, R. and Rand, A. L., 1940, *New Guinea Expedition: Fly River Area, 1936–1937*. Robert M. McBride, New York.

Barrau, J., 1958, Subsistence agriculture in Melanesia. *Bishop Museum Bulletin*, 219, Honolulu.

———— 1959, The sago palms and other food plants of marsh dwellers in the Pacific Islands. *Economic Botany*, 13: 151–162.

———— 1961, Subsistence agriculture in Polynesia and Micronesia. *Bishop Museum Bulletin*, 223, Honolulu.

———— 1968, L'humide et le sec: an essay on ethnobotanical adaptation to contrastive environments in the Indo-Pacific area. In A. P. Vayda (ed), *Peoples and Cultures of the Pacific*: 113–132. Natural History Press, Garden City (N.Y.).

Bayliss-Smith, T. P., 1977, Human ecology and island populations: the problems of change. In T. P. Bayliss-Smith and R. G. Feachem (eds), *Subsistence and Survival: Rural Ecology in the Pacific*: 11–20. Academic Press, London.

Beaver, W. N., 1920, *Unexplored New Guinea*. Seeley, Service & Co., Edinburgh.

Blake, D. H., 1971, Geology and geomorphology of the Morehead-Kiunga area. In *Land Resources of the Morehead-Kiunga Area, Territory of Papua and New Guinea* (*Land Research Series*

* written in Japanese

No. 29): 56–68. Commonwealth Scientific and Industrial Research Organization, Australia, Melbourne.

Bleeker, P., 1971, Soils of the Morehead-Kiunga area. In *Land Resources of the Morehead-Kiunga Area, Territory of Papua and New Guinea (Land Research Series No. 29)*: 69–87. Commonwealth Scientific and Industrial Research Organization, Australia, Melbourne.

Boserup, E., 1965, *The Economics of Agrarian Change under Population Pressure*. Aldine, Chicago.

——— 1976, Environment, population and technology in primitive societies. *Population and Development Review*, 2: 21–36.

Bronson, B., 1977, The earliest farming: demography as cause and consequence. In C. A. Reed (ed), *Origins of Agriculture*: 23–48. Mouton, The Hague.

Brookfield, H. C. and Hart, D., 1971, *Melanesia: A Geographical Interpretation of an Island World*. Methuen, London.

Bulmer, R., 1969, The strategies of hunting in New Guinea. *Oceania*, 38: 302–318.

Bulmer, S., 1966, Pig-bone from two archaeological sites in the New Guinea Highland. *Journal of Polynesian Society*, 75: 504–505.

Burkill, I. H., 1960, The organography and the evolution of Dioscoreaceae, the family of yams. *Journal of the Linnean Society of London (Botany)*, 56: 319–412.

Carneiro, R. L. 1957, *Subsistence and Social Structure*. Ph.D. Dissertation, University of Michigan.

Carpenter, C. R., 1934, A field study of the behavior and social relations of howling monkeys. *Comparative Psychology Monographs*, 10: 1–168.

Casteel, R. W., 1972–73, Two static maximum population-density models for hunter-gatherers: a first approximation. *World Archaeology*, 4: 19–40.

Chowning, A., 1963, Proto-Melanesian plant names. In J. Barrau (ed), *Plants and the Migrations of Pacific Peoples*: 39–44. Bishop Museum Press, Honolulu.

Clarke, W. C., 1971, *Place and People: An Ecology of a New Guinea Community*. Australian National University Press, Canberra.

——— 1976, Maintenance of agriculture and human habitats within the tropical forest ecosystem. *Human Ecology*, 4: 247–259.

Conklin, H. C., 1957, *Hanunóo Agriculture*. Food and Agriculture Organization, Rome.

Conroy, W. L. and Bridgland, L. A., 1947, Native agriculture in Papua-New Guinea. In E. H. Hipsley and F. W. Clements (eds), *Report of the New Guinea Nutritional Survey Expedition:* 72–91. Department of External Territories, Australia, Canberra.

Corden, M. W., 1970, Some observations on village life in New Guinea. *Australian Institute of Anatomy, Food and Nutrition, Notes and Review*, 27: 77–82.

Corner, E.J.H., 1966, *The Natural History of Palms*. Weidenfeld and Nicolson, London.

Crocombe, R. G., 1972, Land tenure in the South Pacific. In R. G. Ward (ed), *Man in the Pacific Islands:* 219–251. Clarendon, Oxford.

Crow, J. F., 1958, Some possibilities for measuring selection intensities in man. *Human Biology*, 30: 1–13.

Digby, A., 1949, Technique and the time factor in relation to economic organization. *Man*, 49: 16–18.

Dobzhansky, T., 1968, Adaptedness and fitness. In R. C. Lewontin (ed), *Population Biology and Evolution:* 109–121. Syracuse University Press, New York.

Dyke, B. and MacCluer, J. W., 1973, Estimation of vital rates by means of Monte Carlo simulation. *Demography*, 10: 383–403.

Erasmus, C. J., 1955, Work pattern in a Mayo village. *American Anthropologist*, 57: 322–333.

Firth, R., 1939, *Primitive Polynesian Economy*. Routledge, London.

———— 1946, *Malay Fishermen: Their Peasant Economy*. Routledge and Kegan Paul, London.

Fountain, O. C., 1966, *Wulkum: Land, Livelihood and Change in a New Guinea Village*. M. A. Thesis, Victoria University of Wellington, New Zealand.

Gayden, E. L., 1974, Transportation and communication systems. In F. Sargent II (ed), *Human Ecology:* 235–252. North-Holland, Amsterdam.

Golson, J., 1971, Both sides of the Wallace Line: Australia, New Guinea, and Asian prehistory. *Archaeology and Physical Anthropology in Oceania*, 6: 124–144.

———— 1972, Land connections, sea barriers and the relationship of Australian and New Guinea prehistory. In D. Walker (ed), *Bridge and Barrier: The Natural and Cultural History of Torres Strait:* 375–397. Australian National University, Canberra.

——— 1977, No room at the top: agricultural intensification in the New Guinea highlands. In J. Allen, J. Golson and R. Jones (eds), *Sunda and Sahul: Prehistoric Studies in Southeast Asia, Melanesia and Australia:* 601–638. Academic Press, London.

Gross, D. R., 1975, Protein capture and cultural development in the Amazon Basin. *American Anthropologist*, 77: 526–874.

Haberland, E. and Seyfarth, S., 1974, Die Yimar am Oberen Korowori (Neuguinea). *Studien zur Kulturkunde*, 36. Franz Steiner, Wiesbaden.

Hambridge, K. M., Hambridge, C., Jacobs, M. and Baum, J. B., 1972, Low level of zinc in hair, anorexia, poor growth, and hypogeusia in children. *Pediatric Research*, 6: 868–874.

Harris, D. R., 1977, Subsistence strategies across Torres Strait. In J. Allen, J. Golson and R. Jones (eds), *Sunda and Sahul: Prehistoric Studies in Southeast Asia, Melanesia and Australia:* 421–463. Academic Press, London.

Hassan, F. A., 1973, On mechanisms of population growth during the Neolithic. *Current Anthropology*, 14: 535–540.

——— 1981, Demographic Archaeology. Academic Press, New York.

Hayden, B., 1975, The carrying capacity dilemma: an alternate approach. *American Antiquity*, 40: 11–21.

Hodges, K., Fysh, C. F. and Rienits, K. G., 1947, New Guinea and Papuan food composition tables. In E. H. Hipsley and F. W. Clements (eds), *Reports of the New Guinea Nutrition Survey Expedition:* 273–279. Department of External Territories, Australia, Canberra.

Hogbin, H. I., 1938–39, Tillage and collection: a New Guinea economy. *Oceania*, 9: 127–151, 286–325.

——— 1970, A New Guinea childhood: from weaning till the eighth year in Wogeo. In J. Middleton (ed), *From Child to Adult: Studies in the Anthropology of Education:* 134–162. University of Texas Press, Austin.

Holmes, L. D., 1979, The function of kava in modern Samoan culture. In D. H. Efron, B. Homstedt and N. S. Kline (eds), *Ethnopharmacologic Search for Psychoactive Drugs:* 107–118.

Raven Press, New York.
Irimoto, T., 1973, Daily space use pattern of male breath-hold abalone divers. *Journal of Human Ergology*, 2: 59–74.
Johnson, A., 1975, Time allocation in a Machiguenga community. *Ethnology*, 14: 301–310.
Jones, R., 1980, Hunters in the Australian coastal savanna. In D. R. Harris (ed), *Human Ecology in Savanna Environments:* 107–146, Academic Press, London.
Kemp, W. B., 1971, The flow of energy in a hunting society. *Scientific American*, 224: 104–115.
Kimura, N., 1979*, Pests of sago palm and their control. *Japanese Journal of Tropical Agriculture*, 23: 142–148.
Kobayashi, K., 1969, Changing patterns of differential fertility in the population of Japan. *Proceedings of the Eighth International Congress of Anthropological and Ethnological Sciences*, 345–347, Tokyo.
Kormondy, E. J., 1974, Natural to human ecosystems. In F. Sargent II (ed), *Human Ecology:* 27–43. North-Holland, Amsterdam.
Kroeber, T., 1964, *Ishi: Last of His Tribe.* Parnassus, Berkeley.
Kuchikura, Y., 1975, The structure of the fishing activity system and division of labor according to age in a local fishing community of Okinawa. *Journal of Human Ergology*, 3: 105–130.
Landtman, G., 1927, *The Kiwai Papuans of British New Guinea.* Methuen, London.
Laughlin, W. S., 1968, Hunting: an integrating biobehavior system and its evolutionary importance. In R. B. Lee and I. DeVore (eds), *Man the Hunter:* 304–320. Aldine, Chicago.
Lea, D.A.M., 1964, *Abelam Land and Sustenance.* Ph.D. Dissertation, Australian National University, Canberra.
——— 1972, Indigenous horticulture in Melanesia. In R. G. Ward (ed), *Man in the Pacific Islands:* 252–279. Clarendon, London.
Lee, R. B., 1968, What hunters do for a living, or, how to make out on scarce resources. In R. B. Lee and I. DeVore (eds), *Man the Hunter:* 30–48. Aldine, Chicago.
Lee, R. B. and DeVore, I., 1968, Problems in the study of hunters and gatherers. In R. B. Lee and I. DeVore (eds), *Man the Hunter:* 3–12. Aldine, Chicago.
Littlewood, R. A., 1972, *Physical Anthropology of the Eastern*

Highlands of New Guinea. University of Washington Press, Seattle.

Lotka, A. J., 1925, On the true rate of natural increase. *Journal of the American Statistical Association*, 20: 305–339.

MacCluer, J. W., Neel, J. V. and Chagnon, N. A., 1971, Demographic structure of a primitive population: a simulation. *American Journal of Physical Anthropology*, 35: 193–207.

Malcolm, L. A., 1970, Growth and Development in New Guinea—A Study of the Bundi People of the Madang District. *Institute of Human Biology Papua-New Guinea Monograph Series*, No. 1. Institute of Human Biology, Madang.

Malinowski, B., 1922, *Argonauts of Western Pacific*. Routledge, London.

Massal, E. and Barrau, J., 1956, Food plant in the south sea islands. *South Pacific Commission Technical Paper*, 94, Noumea.

McAlpine, J. R., 1971, Climate of the Morehead-Kiunga area. In *Land Resources of the Morehead-Kiunga area, Territory of Papua and New Guinea (Land Research Series No. 29)*: 46–55. Commonwealth Scientific and Industrial Research Organization, Australia, Melbourne.

McArthur, M., 1972, Food. In P. Ryan (ed), *Encyclopaedia of Papua and New Guinea*: 433–447. Melbourne University Press, Melbourne.

McCance, R. A. and Widdowson, E. M., 1960, *The Composition of Foods (Medical Research Council Special Series No. 297)*. Her Majesty's Stationary Press, London.

McCarthy, F. D. and McArthur, M., 1960, The food quest and the time factor in aboriginal economic life. In C. P. Mountford (ed), *Records of the American-Australian Scientific Expedition to Arnhem Land, Vol. 2: Anthropology and Nutrition*: 145–194. Melbourne University Press, Melbourne.

McCay, B. J., 1978, System ecology, people ecology, and the anthropology of fishing communities. *Human Ecology*, 6: 397–422.

Mead, M., 1975, *Growing up in New Guinea*. William Morrow, New York.

Minge-Klevana, W., 1980, Does labor time decrease with industrialization? a survey of time-allocation studies. *Current Anthropology*, 21: 279–298.

Miyake, I. and Yamamoto, K., 1976*, *SPSS Statistical Package (I) Volume for Fundamentals*. Toyokeizai-shinsha, Tokyo.

Miyake, I., Nakano, Y., Mizuno, K. and Yamamoto, K., 1976*, *SPSS Statistical Package (II) Volume for Analysis*. Toyo-keizai-shinsha, Tokyo.
Morren, G.E.B.Jr., 1975, *Settlement Strategies and Hunting in a New Guinea Society*. Ph.D. Dissertation, Columbia University.
Morris, H. S., 1953, *Report on a Melanau Sago Producing Community in Sarawak (Colonial Research Studies No. 9)*. Colonial Office, London.
Murdock, G. P., 1949, *Social Structure*. Macmillan, New York.
Nag, M., 1962, Factors affecting human fertility in nonindustrial societies: a cross-cultural study. *Yale University Publications in Anthropology*, No. 66.
Nag, M., White, B. and Peet, R. C., 1978, An anthropological approach to the study of the economic value of children in Java and Nepal. *Current Anthropology*, 19: 293–306.
Nakao, S., 1967*, On the origin of agriculture. In S. Morishita and T. Kira (eds), *Nature: Ecological Studies*. Chuokoron, Tokyo.
National Statistical Office (Papua New Guinea), 1981, 1980 National Population Census Preliminary Bulletin No. 1: Field Counts—All Provinces and Districts. Port Moresby.
Nie, N. H., Hull, C. H., Jenkins, J. G., Steinbrenner, K. and Bent, D. H., 1975, *Statistical Package for the Social Sciences (2nd Edition)*. McGraw-Hill, New York.
Nishikawa, G., Oda, K. and Suganuma, H., 1979*, Botanical description of sago palm. *Japanese Journal of Tropical Agriculture*, 23: 123–130.
Odum, E. P., 1971, *Fundamentals of Ecology (3rd Edition)*. W. B. Saunders Company, Philadelphia.
Ohtsuka, R., 1970, Ecology of Nasake fishermen: the temporal and spatial structure of their activity system as revealed by an individual-tracing method. *Journal of Anthropological Society of Nippon*, 78: 121–139.
——— 1971, Microclimatic conditions under which the Nasake fishermen are working at sea in summer. *Journal of Anthropological Society of Nippon*, 79: 1–8.
——— 1972, Fishing time and catches among hand-line fishermen with special reference to individual variation and the relationship between fishing activity and natural environment. *Journal of Human Ergology*, 1: 3–18.

——— 1975*, Hunting strategies and skill of Papuans. *Popular Medicine*, 64: 120–124.

——— 1977a, The sago eaters: an ecological discussion with special reference to the Oriomo Papuans. In J. Allen, J. Golson and R. Jones (eds), *Sunda and Sahul: Prehistoric Studies in Southeast Asia, Melanesia and Australia:* 465–492. Academic Press, London.

——— 1977b, Individual variations in the fishing activity of Nasake Islanders. In H. Watanabe (ed), *Human Activity System: Its Spatiotemporal Structure:* 41–59. University of Tokyo Press, Tokyo.

——— 1977c, Time-space use of the Papuans depending on sago and game. In H. Watanabe (ed), *Human Activity System: Its Spatiotemporal Structure:* 231–260. University of Tokyo Press, Tokyo.

——— 1978*, The lowland Papuans depending on hunting. *Anima*, 59: 81–87.

——— 1979, Behavioral skill of shellfish divers in an Okinawan community. *Proceedings of 9th Asian Conference on Occupational Health*, 158–162, Seoul.

Ohtsuka, R. and Suzuki, T., 1978, Zinc, copper and mercury in Oriomo Papuans' hair. *Ecology of Food and Nutrition*, 6: 243–249.

Oomen, H.A.P.C., 1971, Ecology of human nutrition in New Guinea—evaluation of subsistence patterns. *Ecology of Food and Nutrition*, 1: 1–16.

Paijmans, K., 1971, Vegetation, forest resources, and ecology of the Morehead-Kiunga area. In *Land Resources of the Morehead-Kiunga Area, Territory of Papua and New Guinea (Land Research Series No. 29):* 88–113. Commonwealth Scientific and Industrial Research Organization, Australia, Melbourne.

Peters, F. E., 1957, Chemical composition of South Pacific foods—an annotated bibliography. *South Pacific Commission Technical Paper*, 100, Noumea.

Peterson, J. T. and Peterson, W., 1977, Implications of contemporary and prehistoric exchange system. In J. Allen, J. Golson and R. Jones (eds), *Sunda and Sahul: Prehistoric Studies in Southeast Asia, Melanesia and Australia:* 533–564. Academic Press, London.

Polunin, I. and Saunders, M., 1958, Infertility and depopula-

tion: a study of the Murut tribes of north Borneo. *The Lancet*, 2: 1005–1008.
Pope, S. T., 1918, Yahi archery. *University of California Publications for American Archaeology and Ethnology*, 13: 103–152.
Powell, J. M., 1976, Ethnobotany. In K. Paijmans (ed), *New Guinea Vegetation:* 106–183. Australian National University Press, Canberra.
Purseglove, J. W., 1978, *Tropical Crops: Monocotyledons (2nd Edition)*. English Language Book Society and Longman, London.
Rappaport, R. A., 1968, *Pigs for the Ancestors: Ritual in the Ecology of a New Guinea People.* Yale University Press, New Haven.
Ray, S. H., 1907, *Report of the Cambridge Anthropological Expedition to Torres Strait. Volume 3: Linguistics.* Cambridge University Press, Cambridge.
Reinman, F. M., 1967, Fishing: an aspect of Oceanic economy. an archaeological approach. *Fieldiana Anthropology*, 56: 99–208.
Richards, A. I., 1939, *Land, Labour and Diet in Northern Rhodesia (2nd Edition)*. Oxford University Press, London.
Ruddle, K., Johnson, D., Townsend, P. K. and Rees, J. D., 1978, *Palm Sago: A Tropical Starch from Marginal Lands.* East-West Center, Honolulu.
Sahlins, M. D., 1965, On the sociology of primitive exchange. In M. Banton (ed), *The Relevance of Models for Social Anthropology (ASA Monographs No. 1)*. Tavistock, London.
——— 1968, Notes on the original affluent society. In R. B. Lee and I. DeVore (eds), *Man the Hunter:* 85–89. Aldine, Chicago.
——— 1972, *Stone Age Economics.* Tavistock, London.
Sasaki, N., Takemori, K., Ohtsuka, R. and Suzuki, T., 1981, Mineral contents in hair from Oriomo Papuans and Akita dwellers. *Ecology of Food and Nutrition*, 11: 117–120.
Schaller, G. B., 1963, *The Mountain Gorilla: Ecology and Behavior.* University of Chicago Press, Chicago.
Service, E. R., 1966, *The Hunters.* Prentice-Hall, Englewood Cliffs.
Sever, L. E., 1975, Zinc and human development: a review. *Human Ecology*, 3: 43–57.
Skeldon, R., 1979, Introduction to the demography of Papua

New Guinea. In R. Skeldon (ed), The Demography of Papua New Guinea. *IASER* (*Institute of Applied Social and Economic Research, Papua New Guinea*) *Monograph*, 11: 1–9.

Spencer, J. E., 1963, The migration of rice from mainland Southeast Asia into Indonesia. In J. Barrau (ed), *Plants and the Migrations of Pacific Peoples:* 83–89. Bishop Museum Press, Honolulu.

Stauder, J., 1971, *The Majangir: Ecology and Society of a Southern Ethiopian People.* Cambridge University Press, Cambridge.

Street, J. M., 1969, An evaluation of the concept of carrying capacity. *The Professional Geographer*, 21: 104–107.

Suzuki, T., 1970*, Biographical introduction to human ecology. In H. Katsunuma and T. Suzuki (eds), *Human Ecology Notes:* 209–243. University of Tokyo Press, Tokyo.

——— 1977, Human ecology in understanding environmental health problem. *Reviews on Environmental Health*, 2: 204–219.

——— 1980*, *Methodology of Human Ecology.* University of Tokyo Press, Tokyo.

Takeda, J., 1972, An ecological study of bear-hunting activities of the Matagi, Japanese traditional hunters. *Journal of Human Ergology*, 1: 167–187.

Tanaka, J., 1976, Subsistence ecology of Central Kalahari San. In R. B. Lee and I. Devore (eds), *Kalahari Hunter-Gatherers: Studies of the !Kung San and Their Neighbors:* 98–119. Harvard University Press, Cambridge.

——— 1980, *The San: Hunter-Gatherers of the Kalahari.* University of Tokyo Press, Tokyo.

Tax, S., 1963, *Penny Capitalism.* University of Chicago Press, Chicago.

Thomson, B. P., 1954, *Two Studies in African Nutrition.* Manchester University Press, Manchester.

Townsend, P. K., 1969, *Subsistence and Social Organization in a New Guinea Society.* Ph.D. Dissertation, University of Michigan.

——— 1971, New Guinea sago gatherers: a study of demography in relation to subsistence. *Ecology of Food and Nutrition*, 1: 19–24.

——— 1974, Sago production in a New Guinea economy. *Human Ecology*, 2: 217–236.

Townsend, W. H., 1969, Stone and steel tool use in a New

Guinea society. *Ethnology*, 8: 199–205.
Ulijaszek, S. and Poraituk, S., 1981, The sago subsistence of the people of the Purari River delta. *Purari River (Wabo) Hydroelectric Scheme Environmental Studies*, Vol. 19, Office of Environment and Conservation, Central Government Offices (Papua New Guinea), Waigani (Port Moresby).
Vayda, A. P. and McCay, B. J., 1975, New directions in ecology and ecological anthropology. *Annual Review of Anthropology*, 4: 293–306.
Voorhoeve, C. L., 1968, The Central and South New Guinea Phylum. *Pacific Linguistics*, A. 16: 1–17.
——— 1970, Some notes on the Suki-Gogodala subgroups of the Central and South New Guinea Phylum. *Pacific Linguistics*, C. 13: 1246–1270.
Walker, D. [ed], 1972, *Bridge and Barrier: The Natural and Cultural History of Torres Strait*. Australian National University, Canberra.
Walsh, R. J., 1972, Physical anthropology. In P. Ryan (ed), *Encyclopaedia of Papua and New Guinea:* 896–901. Melbourne University Press, Melbourne.
Ward, R. G. and Lea, D.A.M., 1970, *An Atlas of Papua and New Guinea*. Collins and Longman, London.
Watanabe, H., 1964, The Ainu: a study of ecology and the system of social solidarity between man and nature in relation to group structure. *Journal of Faculty of Science, University of Tokyo*, V (2): 1–164.
——— 1971, Periglacial ecology and the emergences of *Homo sapiens*. In *The Origin of Homo sapiens (Ecology and Conservation):* 271–285. UNESCO, Paris.
——— 1975, Bow and arrow census in a west Papuan lowland community: a new field for functional-ecological study. *Occassional Papers in Anthropology (University of Queensland)*, *5*, St. Lucia.
——— 1977a [ed] *Human Activity System: Its Spatiotemporal Structure*. University of Tokyo Press, Tokyo.
——— 1977b*, Introduction to ecological anthropology. In H. Watanabe (ed), *Series of Anthropology 12: Ecology:* 3–29. Yuzankaku, Tokyo.
——— 1978*, Origin of tools and hominization. *Kagaku*, 48: 221–234.
White, J. P., 1971a, New Guinea and Australian prehistory:

the 'neolithic' problem. In D. J. Mulvaney and J. Golson (eds), *Aboriginal Man and Environment in Australia:* 182–195. Australian National University Press, Canberra.

——— 1971b, New Guinea: the first phase in Oceanic settlement. In R. C. Green and M. Kelly (eds), *Studies in Oceanic Culture History Volume 2:* 45–66. Bishop Museum, Honolulu.

——— 1972, Ol Tumbuna: archaeological excavation in the eastern central highlands, Papua New Guinea. Department of Prehistory, Research School of Pacific Studies, Australian National University, *Terra Australis*, 2.

Williams, F. E., 1936, *Papuans of the Trans-Fly*. Clarendon, Oxford.

Wollaston, A.F.R., 1912, *Pygmies and Papuans*. Smith, Elder, London.

Wurm, S. A., 1971a, The Papuan linguistic situation. In I. A. Sebeck (ed), *Current Trends in Linguistics Volume 8: Linguistics in Oceania*. Mouton, The Hague.

——— 1971b, Notes on the linguistic situation in the Trans-Fly area. *Papers in New Guinea Linguistics*, 14: 115–172.

Yen, D. E., 1971, The development of agriculture in Oceania. In R. C. Green and M. Kelly (eds), *Studies in Oceanic Culture History Volume 2:* 1–12. Bishop Museum, Honolulu.

——— 1974, Arboriculture in the subsistence of Santa Cruz, Solomon Islands. *Economic Botany*, 28: 247–284.

Index